COMPLETE
BOOK OF
DWARF
CICHLIDS

By Hans-Joachim Richter

Translator: William Charlton

Microgeophagus altispinosa.

All photos other than product photos by the author unless specifically noted otherwise.

Distributed in the UNITED STATES by T.F.H. Publications, Inc., One T.F.H. Plaza, Neptune City, NJ 07753; in CANADA to the Pet Trade by H & L Pet Supplies Inc., 27 Kingston Crescent, Kitchener, Ontario N2B 2T6; Rolf C. Hagen Ltd., 3225 Sartelon Street, Montreal 382 Quebec; in CANADA to the Book Trade by Macmillan of Canada (A Division of Canada Publishing Corporation), 164 Commander Boulevard, Agincourt, Ontario M1S 3C7; in ENGLAND by T.F.H. Publications Limited, Cliveden House/Priors Way/Bray, Maidenhead, Berkshire SL6 2HP, England; in AUSTRALIA AND THE SOUTH PACIFIC by T.F.H. (Australia) Pty. Ltd., Box 149, Brookvale 2100 N.S.W., Australia; in NEW ZEALAND by Ross Haines & Son, Ltd., 18 Monmouth Street, Grey Lynn, Auckland 2, New Zealand; in SINGAPORE AND MALAYSIA by MPH Distributors (S) Pte., Ltd., 601 Sims Drive, #03/07/21, Singapore 1438; in the PHILIPPINES by Bio-Research, 5 Lippay Street, San Lorenzo Village, Makati Rizal; in SOUTH AFRICA by Multipet Pty. Ltd., 30 Turners Avenue, Durban 4001. Published by T.F.H. Publications, Inc. Manufactured in the United States of America by T.F.H. Publications, Inc.

TABLE OF
CONTENTS

There are many ways in which cichlids spawn and this book will inform you about most of the ways. The photo above shows a male and female mouthbrooder, *Haplochromis burtoni*, which is borderline between fully-grown cichlids and the dwarf varieties covered in this book. The male is much more colorful than the female. The female carries the eggs (in some mouthbrooding species, the males may carry the eggs) and has a color which renders her camouflaged. The male has egg spots in his anal fin. This is to attract the female to snap at them during spawning, thus fertilizing the eggs in her mouth. It's all theory since there are many successful mouthbrooders which do not have egg spots on the anal fin.

The diversity of species of exotic fishes available in the aquarium hobby has increased to such an extent in the last fifty years that the choice can be difficult even for experienced aquarists. The motives for buying particular fishes and devoting one's attention to their keeping and breeding are manifold in nature. In his choice, many a fancier allows himself to be influenced by the color of a fish, while others are influenced by form. But, just as easily, ease of keeping or the enjoyment of breeding delicate species, and perhaps also the rarity of a species, can play a role in the decision.

Cichlids, ever since they became fixtures in our aquaria, have filled many aquarists with enthusiasm. Their active behavior in the defense of their territories, their courtship behavior, and the parental brood-care behavior offer the observer fascinating variety. There also are numerous species that, on the basis of their splendid coloration alone, have developed in aquarists the wish to keep them some day. In many cases, however, the size of these fishes and their habits are in conflict with our ideas of a decorative tropical fish aquarium. For this reason, one seldom finds beautiful aquaria with larger cichlids. One has far fewer difficulties in a creative sense as well as in the choice of tank size with smaller cichlids, which because of their size are not able to destroy the plants and the furnishings of the aquarium. We designate as dwarf cichlids those cichlids that do not grow much longer than ten centimeters, as well as those that, because of their minimal body depth, still produce the impression of being small. The behavior patterns of dwarf cichlids are generally similar to those of their larger relatives, but they are comparatively peaceful fishes that one can also keep successfully with other fishes in a community tank.

I have kept and spawned in the aquarium the majority of the dwarf cichlid species described in this book. The experiences I have gained from this form the foundation of this book. It should provide advice for the successful keeping and spawning of this group of fishes, and should make the slow process of searching through the voluminous literature easier for those who occupy themselves with specific species. Finally, it might attract more aquarists to these beautiful and interesting fishes.

It is to be hoped that the abundance of photographs and the descriptions of the individual species will be of assistance to many fanciers in the selection of the dwarf cichlids that are appropriate for them.

I do not wish to fail to thank all those who supported me in obtaining many cichlid species as well as the necessary literature, or contributed their advice to the creation of this book. I thank Neumann Verlag for its cooperation and constructive collaboration.

Hans Joachim Richter

Diversity reaches beyond the realm of imagination when you consider how this new *Lamprologus* species spawns. The female lays her eggs in an empty snail shell. Supposedly, it prefers the empty shell of a single species (in nature) but in the aquarium it usually settles for whatever is available. The eggs are fertilized and hatch in the safety of the empty snail shell. No one has, to date, ever been able to spawn these fish in anything but an empty snail shell.

This magnificent male *Apistogramma agassizi* shows the magnificent brilliance that dwarf cichlids can attain. This photo was used as the cover of the original German edition.

General Section

Pre-spawning behavior of a typical mouthbrooding cichlid, *Haplochromis burtoni*. In Germany, this species is not considered as a dwarf cichlid. In America, England and Australia it is considered small enough.

Aequidens dorsiger displaying typical pre-spawning behavior. The female is the more colorful of the pair. She is ready to spawn as is indicated by the ovipositor extending from her anal pore. This behavior sometimes leads to the pair locking jaws or "kissing."

Over the course of decades, the concept of the dwarf cichlid has become a collective term in the aquarium hobby for all cichlids that, as a rule, are less than ten centimeters long. Thus, it is not a question of a group of fishes that are classified together systematically on the basis of common characters, such as is the case with cichlids as a whole, but rather of a concept used in the aquarium hobby to describe only those smaller species within the large group of cichlids (in which species of up to a half meter in length are found) that are suitable for keeping in the aquarium.

Cichlids possess only one nostril on each side of the head, and on this basis are readily distinguished from similar perchlike fishes. In contrast to this, all other freshwater fishes are provided with two nostrils, an inhalent one and an exhalent one. Unlike many other perchlike fishes that generally possess two dorsal fins, cichlids have only a single dorsal fin.

In 1908, Engmann used the designation dwarf cichlids for the first time, yet he did not go into the details of the provenance of this concept. In an essay he reported on *Geophagus taeniatus* (= *Apistogramma taeniata*) and *Geophagus agassizii* (= *Apistogramma agassizii*). In subsequent literature, *Apistogramma agassizii* is expressly designated by the popular name dwarf cichlid. At this time *Nannacara anomala* received the German name Gestreifter Zwergbuntbarsch (= Striped Dwarf Cichlid). Later, all *Apistogramma* species were referred to by the root words dwarf cichlid: For example, Yellow Dwarf Cichlid (*Apistogramma borellii*), Prettyfin Dwarf Cichlid (*Apistogramma ornatipinnis*), Amazon Dwarf Cichlid (*Apistogramma pertensis*). Subsequently, other newly discovered small cichlids also were assigned to this group more and more frequently.

Like the majority of cichlids, the small species are also principally inhabitants of the lower water levels. They generally lead a hidden life just above the bottom. The majority of species are therefore also hidden spawners, the living spaces of which are the riparian zones of the bodies of water in which they occur.

With the exception of the open-spawning species, we find a noticeable sexual dimorphism in dwarf cichlids. Whereas males are particularly conspicuous in form and coloration, particularly during courtship, females usually exhibit an inconspicuous camouflage coloration that allows them to blend into their living space. During the period of brood care, however, females of the genus *Apistogramma* often are a conspicuous yellow color; any markings possibly present beforehand have almost completely disappeared, but in their place well-defined patterns of black markings become very prominent. Distinguishing the sex is quite easy in many dwarf cichlids. With species from Lake Tanganyika and certain open-spawning species, however, it takes a practiced eye to determine the sex.

Even though the word "dwarf" is associated with the idea of something very small, today some authors no longer see this nearly so strictly and include fishes that attain a length even of up to 15 centimeters in the group called dwarf cichlids. In comparison to the large species, however, these fish are still truly dwarfs.

Apistogramma cacatuoides; the male dances before the female in a typical pre-nuptial display. As spawning time approaches, the male's colors become more intense.

Note: The subdivision of the dwarf cichlids in this book is made in consideration of their spawning behavior. Within this arrangement the genera and, within the genera, the species are described in alphabetical order.

The life on our planet is of a diversity of forms that is scarcely conceivable to the average person. Each of us is familiar with a portion of the animal and plant world by particular names. The German, English, French, and Spanish names or the names from any other living language for a particular animal or plant are called popular names. At the same time, it also happens that species often are furnished with several popular names. An example among the dwarf cichlids is the Cockatoo Dwarf Cichlid, which is also called the Indian Cichlid. The popular names certainly are completely adequate for everyday speech, but with these names it is impossible to address a species in precise, unambiguous terms that are understandable throughout the world. New organisms are discovered daily and must be classified in the nomenclature. The basis for this classification is the rules of an internationally valid scientific nomenclature and systematics. Systematics considers kindred relationships that have been identified by science.

Every species is unequivocally differentiated from every other species by means of a scientific name. The designation of a species with a scientific name takes place according to the International Rules for Zoological Nomenclature. These were originally adopted in 1905 at the International Congress of Zoologists in Paris and are still valid today after several modifications.

The species is identified by a generic and a specific name according to binomial nomenclature. The name of the describer of the species and the year of the publication of the description are appended. Thus, a complete scientific name reads as follows: *Apistogramma cacatuoides* Hoedeman, 1951. This signifies a fish of the genus *Apistogramma* with the specific name *cacatuoides* (like a cockatoo). This species was described by Hoedeman in 1951.

If the scientific designation consists of three names, this indicates that the described fish belongs to a species of which subspecies have been described; for example: *Neolamprologus leleupi longior* (Staeck, 1980). This means that in addition to the subspecies mentioned, there also must be at least one other subspecies, namely the nominate subspecies *Neolamprologus leleupi leleupi* (Poll, 1956), usually written *Neolamprologus l. leleupi*.

The determination of a subspecies is a very difficult taxonomic problem and is often dealt with very irresponsibly by numerous first describers for reasons of personal ambition. Fundamentally, color differences alone should not come into question for proposing a new subspecies. For example, on account of the vast area of occurrence of *Apistogramma agassizii*, there are numerous varieties with differences in coloration and markings, but these differences are not sufficient to be able to treat them as separate subspecies. In this connection, it is clearly only a question of color varieties of different populations.

Striking differences, such as in build and behavior, particularly in spawning and brood-care behavior, possibly justify taking particular species from a genus and establishing a new genus for these species. Thus, because of differences in build and behavior, the species *Apistogramma ramirezi* was taken from the genus *Apistogramma* and a new genus was established for it, namely *Microgeophagus* Axelrod, 1971. The case is much the same with *Pseudocrenilabrus multicolor* (Hilgendorf, 1903), the Multicolored Mouthbrooder, which was previously classified in the genera *Haplochromis* and *Hemihaplochromis*. According to the Rules of Zoological Nomenclature, in such cases the name of the first describer and the date of first publication are placed in parentheses. Ramirez's Dwarf Cichlid was named *Apistogramma ramirezi* Myers and Harry, 1948, by its first describers. After the new genus was established, the scientific name was changed to *Microgeophagus ramirezi* (Myers and Harry, 1948). Take note, however, that in each case the specific name is never changed after the first description (rule of priority).

The original description of a species does not always become so well known that double and multiple descriptions are ruled out; in fact, it is almost the rule rather than the exception that a species can be found under various names in the literature. The only valid name is the one that that species received from the first describer and for which the type specimens must exist at institutions of zoology. Every revision of a specific description presumes the examination of the appropriate type specimens. New names for previously described species are called synonyms and are invalid. As they nevertheless frequently appear in place of the valid names in the literature, they are also mentioned in the descriptions of the individual species in this book.

From a systematic point of view, cichlids are a family in the class of bony fishes. In this they are taxonomically classified as follows:

Class	Osteichthyes	bony fishes
Subclass	Actinopterygii	spiny-rayed fishes
Superorder	Teleostei	higher bony fishes
Order	Perciformes	perch-like fishes
Suborder	Percoidei	relatives of perches
Family	Cichlidae	cichlids

The description of a fish species is accomplished according to specific criteria. To these belong size, build, distribution, and behavior. The most important external characters are the number of fin rays and the number of individual scales in the longitudinal and vertical scale rows.

Every fish species has a so-called fin formula. In this the spines or hard rays are indicated by Roman numerals and the soft rays by Arabic numerals. The two values are separated from each other by a slash in the fin formula. With soft rays it can be a question of branched and jointed rays or of unbranched nonjointed or jointed rays. One also separates unbranched soft rays from branched ones by a stroke. The letter preceding the number designates the type of fin:

A (Anal) = anal fin (Pinna analis)
C (Caudal) = caudal fin (Pinna caudalis)
D (Dorsal) = dorsal fin (Pinna dorsalis)
P (Pectoral) = pectoral fin (Pinna pectoralis)
V (Ventral) = ventral fin (Pinna ventralis)

Apistogramma eunotus has 14 to 15 spines and 6 to 7 soft rays in the dorsal fin, 3 spines and 6 to 7 soft rays in the anal fin, 11 to 12 soft rays in the pectoral fin, and 16 soft rays in the caudal fin.

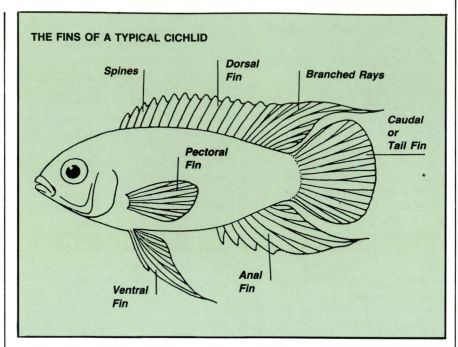

THE FINS OF A TYPICAL CICHLID

Accordingly, the fin formula reads:
D XIV-XV/6–7
A III/6–7
P 11–12
C 16

The scale formula indicates how many scales are counted in the lateral line (Linea lateralis) or, in the event it is lacking, in the median longitudinal row (mLR), from the upper corner of the gill cover as far as the base of the tail fin, and in the vertical row (Linea transversalis = Ltr), as a rule at the point of maximum body depth. In *Apistogramma agassizii* there are 23 scales in the median longitudinal row and 11 to 12 scales in the vertical row. Accordingly, the scale formula is written: mLR 23, Ltr 11–12.

The total length, body length (standard length), head length, maximum body depth, depth of the caudal peduncle, length of the caudal peduncle, snout length, and the description of the body parts are used for further characterization.

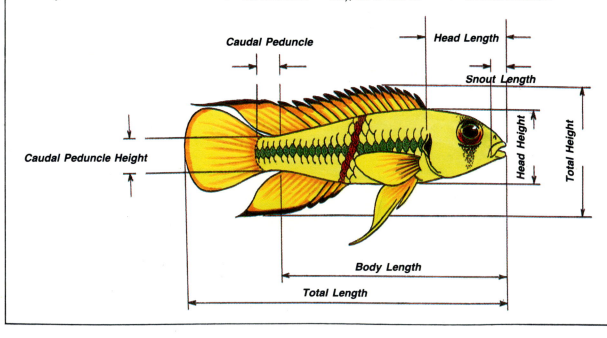

A description of the coloration in cichlids only makes sense, with exceptions, of course, when the possible range in variation is also given. As a rule, in these fishes there are pronounced, mood-dependent colorations that sometimes differ greatly from the normal coloration. Furthermore, age-dependent color varieties are not rare. Mistaken diagnoses, therefore, cannot be ruled out when determination attempts are based solely on a comparison of the coloration.

The body outlines of the various genera should facilitate the initial classification of an unfamiliar cichlid species.

BODY OUTLINES OF THE VARIOUS DWARF CICHLID GENERA COVERED IN THIS BOOK.

Open Spawners

Etroplus

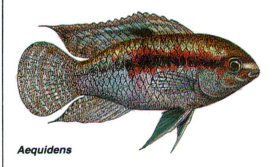

Aequidens

Hemichromis

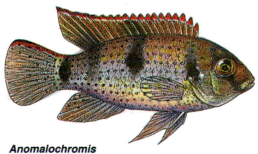

Anomalochromis

Nannacara

Cichlasoma

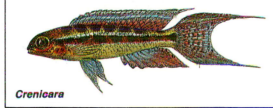

Crenicara

Microgeophagus

Apistogramma

Julidochromis

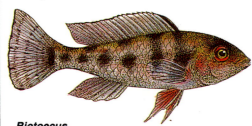

Apistogrammoides

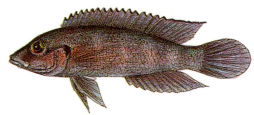

Lamprologus

Biotoecus

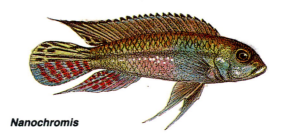

Nanochromis

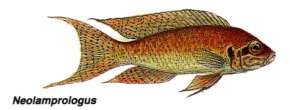

Neolamprologus

Chalinochromis

Cave Spawners

Pelvicachromis

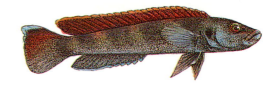

Teleogramma

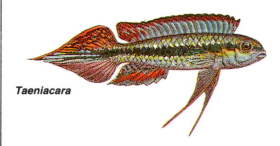

Steatocranus

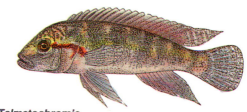

Telmatochromis

Taeniacara

Variabilichromis

Cave Spawners

Chromidotilapia

Labidochromis

Cyprichromis

Melanochromis

Eretmodus

Oreochromis

Iodotropheus

Pseudocrenilabrus

Pseudotropheus

Spathodus

Tanganicodus

Triglachromis

Mouthbrooders

16

Catching exotic fishes in the river basin of the Rio Ucayali in Peru.

The author, Hans Joachim Richter, during a collecting expedition to Peru in 1984.

Attempting to capture *Symphysodon* species, the author circles a fallen tree with a net which reaches from the top of the water to the bottom.

The fallen tree is chopped into small pieces for firewood, and the net is brought closer to shore. The author, being the tallest of the group, is the point man with the net and walks in the deepest part of the small creek.

When the net is brought close to the shore, the overlying vegetation is poked with sticks to drive any fish which took refuge there into the net.

South American Biotope

Dwarf cichlids chiefly occur in South America and Africa. This group of fishes exhibits so many shared characteristics in morphology and behavior that it, in addition to many other factors, can also be cited as proof for the theory of continental drift, according to which South America and Africa were still joined together as a single continent more than 100 million years ago, and have moved apart since that time.

The region from which the majority of the South American species come from is the Amazon basin including the drainage areas of its tributaries. Although this region is incredibly large (about 3,500 by 2,000 kilometers), one can speak of similar environmental conditions almost every place dwarf cichlids are found. This is also true for the other areas of occurrence in South America. In older literature, the Amazon and other large rivers were often cited as places where dwarf cichlids are found. However, it is better to speak of the drainage area of this or that river, since, more precisely, dwarf cichlids are chiefly encountered in smaller, slow-flowing bodies of water, and less commonly in the powerful, often kilometer-wide rivers and streams.

After the rainy season, when extensive areas of the land are flooded (in some regions the water can rise about 15 meters during the rainy season, and the rivers seem boundless), many fishes remain behind outside of their native bodies of water when the water level recedes. In this way they can also reach other rivers and can, under favorable conditions, assuming that a pair is present, establish a new population, which in the course of time diverges more and more from the original population. For example, many color varieties of *Apistogramma agassizii* (Amazon River system) and *Pelvicachromis pulcher* (western Africa) are

known. Many fishes remain behind in isolated bodies of water such as lakes, larger or smaller pools, or so-called residual holes. As the residual holes dry out, one almost exclusively finds larger fishes—the smaller ones having been eaten by the larger ones. Finally, however, all of the fishes in these small bodies of water generally fall prey to predators, such as birds.

The collection of exotic fishes takes place primarily in the dry season, in the larger standing and flowing bodies of water. Collecting is easier at low water, since the

After the net is closed around all the fishes (continued from facing page 18), it is slowly brought to the surface. It contained many species of characins, catfishes and dwarf cichlids . . . but no discusfish, genus *Symphysodon*.

fishes have less opportunity to escape the nets. The water temperatures in the smaller bodies of water, particularly in places that are exposed to direct sunlight for all or part of the day, are comparatively high, whereas they are lower during and after the rainy season. Since exotic fishes are caught only rarely in the rainy season on account of the high water levels, the high water

temperatures measured during the collecting season are often published in the literature without an appropriate remark. This explains why some aquarists keep fishes at water temperatures of up to 30° C, even though 25° C is completely adequate.

If one disregards the three basic types of water (black, white, and clear), the nature of the water differs very little in the various regions. On average, we are dealing with extremely mineral-poor water. The measured conductivity values of 8 to 60 u S in clear and black water are

extremely low and virtually correspond to those of distilled water. Clear water—this water really is clear as glass with a depth of visibility of up to 5 meters—can have a pH of between 4.6 and 6.6. Black water—which, to be sure, generally is also clear, but is a brownish (tea) color—has a pH of between 3.8 and 4.5. The water hardness is also extremely low.

Black-water biotope on the upper course of the Rio Negro.

Biotope on the Yarina Cocha in Peru.

Clear-water biotope south of Manaus.

With black water, maximum values of 0.7° dH in carbonate hardness and of about 1° dH in total hardness have been measured. The majority of the rivers and bodies of water that have water year round have no plant growth at greater depths. Close to shore, the bodies of water often have dense plant growth. These are generally marsh plants that are completely covered with water only at higher water levels. Among rocks and roots as well as broken-off branches one finds a multitude of fishes here, among them dwarf cichlids. As a rule, catching them is not exactly easy, especially since they do not occur in schools, as, for example, many tetra species, and, in addition, the many branches in the water make collecting them much more difficult. For this reason, it is often the case that only a few specimens reach us, at least with initial importations. In the bodies of water of their homelands, dwarf cichlids chiefly feed on tiny shrimps, midge larvae, and insects that fall onto the water's surface. Now and then they even succeed in capturing a small fish.

West African Biotope

On the African continent, two regions are of interest as areas of occurrence of dwarf cichlids:

1. The drainage area of the Congo and its tributaries as well as the river systems in the adjacent countries to the north and northwest as far as Senegal.

2. The large lakes of East Africa.

Like the rain-forest region of the Amazon basin, the territory of tropical West Africa is of immense scale. Innumerable small water courses are present everywhere in the rain forests. Here, depending on the region, one encounters virtually the same conditions as in South America. Because of the heavy human settlement, however, above all near the coast, extensive areas have taken on a savanna-like character in the

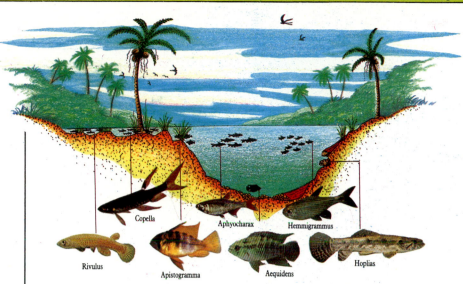

A cross-section of a typical South American stream or igarapé (Brazil) showing the habitats of the various fishes. The dwarf cichlids are represented by the genera *Apistogramma*, which is now actually a *Microgeophagus ramirezi*, and *Aequidens*.

Copella
Aphyocharax
Hemmigrammus
Rivulus
Apistogramma
Aequidens
Hoplias

course of time. Differences in elevation have given rise to waterfalls and fast-flowing bodies of water. In fast-flowing bodies of water, as a rule, live species that are well-adapted to these conditions, such as, for example, *Steatocranus tinanti*. The swim bladder of these fish is developed in such a way that staying on the bottom, not swimming through the water, is the rule.

The slow- to fast-flowing bodies of water in this region generally possess a substrate of fine-grained, light gravel. On and in the substrate one finds, depending on the region, more or less numerous rocks of varying size. In some regions larger rocks and branches lying in the water are covered with large clumps of Congo Water Fern (*Bolbitis heudelotii*). On the banks of the rain-forest streams one often finds, in shallow water as well, vast quantities of Anubias (*Anubias* spp.). Occasionally, the substrate consists of dark volcanic rock. Along the banks, sheltered places under washed out roots, rocks, branches, and overhanging vegetation are present almost everywhere. Dwarf cichlids are collected at these sites in particular. The water often is only 20 to 50 centimeters deep in the places where they are found.

The water parameters in these habitats range from extremely mineral-poor (10 u S) with a pH of 4.5 up to a conductivity of 200 u S and a pH of 7.8. The total hardness lies between 0.5 and 6° dH.

The water generally is clear as glass to slightly brownish. The water temperatures lie between 22° and 27° C, depending on whether the body of water is at a higher elevation and is shaded or whether it is exposed to all-day sunlight and flows in the plain.

East African Biotope

In East Africa, dwarf cichlids almost exclusively occur in the Central African rift valleys and the East African rift valleys, where Lake Malawi (Nyasa) as well as Lake Tanganyika are of central importance. Both lakes are the homelands of many species that have only been discovered for and distributed in the aquarium hobby in the last ten years.

From Lake Tanganyika, the genera *Julidochromis* and *Neolamprologus* are of particular interest to aquarists. With a depth of 1470 meters, Lake Tanganyika is the deepest of the Central African rift valley lakes. It has an area of about 34,000 square kilometers. Its age is estimated to

The author has been elected the expedition's cook. He has to cook what was caught.

The expedition (Peru, 1984) gets under way with a total of eight people and three small boats.

The best fishing is under the trees where it is fairly dark and not too much vegetation (see photo below) is in the way of the nets.

The author is also an expert swimmer, having won swimming medals in Germany. Below: One of the men separates the fishes caught according to behavior.

Pebble shore with reed growth on Lake Tanganyika.

Rocky shore with boulder zone on Lake Tanganyika.

Underwater photograph of a rock zone on Lake Tanganyika.

The rocky shoreline of Lake Malawi where the mbuna (rock-haunting) fishes have their habitat. The rocks on the shore lie in the same approximate position as the rocks under water. The wide white line around the rocks indicates the high water mark. Photo by Dr. Herbert R. Axelrod, from his book *African Cichlids of Lakes Malawi and Tanganyika.*

An aerial view of Monkey Bay, Lake Malawi, where Dr. Axelrod began collecting rift lake cichlids in the late 1940's. The lake has many islands. The islands are surrounded by deep water, trapping the small fishes that are afraid of leaving their secure niches because predacious fish patrol the deeper waters and eat any venturesome small cichlids. This has resulted in many different forms of rock-inhabiting cichlids and created a scientific jumble in the nomenclature of the rift lake species. Photo by Dr. Herbert R. Axelrod.

be about 10 to 15 million years. Along the just under 3,000-kilometer-long shore one finds, besides flat sandy shores, also marshy areas, boulder shores, densely grown regions with a sandy substrate, as well as rock zones. These can take the form of sheer banks, but can also connect with boulder shores. In these places we find dwarf cichlids down to a depth of 30 meters. The water is very clean and clear in this region. The fishes are well adapted to the environment and chiefly inhabit crevices in the rocks and other natural cavities. With an average pH of 8.4 (values of between 7.5 and 9.2 have been measured) and a total hardness of between 7° and 11° dH, as well as a carbonate hardness of about 17° dH, the water has a conductivity of about 600 u S. In contrast to the water chemistry of the West African bodies of water, here we are dealing with relatively mineral-rich, alkaline water. The water temperatures lie between 24° and 28° C. One almost exclusively finds higher aquatic plants only near entering rivers. On the other hand, fairly extensive lawns of algae are encountered chiefly in the boulder and rock zones. Algae represent the dietary basis for many fishes, which eat the algae as well as the microorganisms living in them. With these fishes, the position of the mouth and the nature of the teeth are well adapted to the mode of feeding. The range of visibility under water is astonishing. It can be up to 22 meters in still water. So far just under 200 dwarf cichlid species have been discovered and described from Lake Tanganyika.

Lake Malawi, located farther south in the East African rift valley, is about 600 kilometers long and about 85 kilometers wide. It is only 700 meters deep at its deepest points. On average, however, it is about 270 meters deep. As with Lake Tanganyika, here, too, one finds the most diverse shore forms, where particularly the boulder and rock zones are

An underwater view in Lake Malawi showing the natural habitat of the shoreline species. The rocks seem to have tumbled from the hills (see photo facing page) because no rocks are found in the areas between the hills. In these rock-free areas, water plants grow profusely. Photo by Dr. Herbert R. Axelrod.

In 1949 the mbuna (rock-haunting cichlids of Lake Malawi) were thickly packed on each rocky outcrop. This group of mbuna are grazing on the algae growing on the surface of the rocks. When Dr. Axelrod returned in subsequent trips through the 1980's, he found the population declining due to pollution. Islands which were too far from fishermen were equally depleted, thus putting to rest the "over-fishing" theory of some conservationists. Photo by Dr. Herbert R. Axelrod.

Hemichromis biotope in West Africa.

Pelvicachromis biotope in West Africa.

Boulder shore on Lake Tanganyika.

frequently inhabited by dwarf cichlids. They generally stay in shallow water, where the rocky substrate or the rock walls are covered with heavy growths of algae. The water is very clear; nevertheless, the conductivity is about 220 u S. Lake Malawi is considerably poorer in minerals than is Lake Tanganyika. In it pH values of between 7.7 and 8.8 have been measured. The total hardness varies between 6° and 8° dH. The water temperatures are about the same in both lakes. Even moreso than in Lake Tanganyika, the cichlids in Lake Malawi are adapted to grazing on the large expanses of algae. In the algae they also find vast quantities of different microorganisms. In Lake Malawi, among other cichlids, species of the endemic genera *Melanochromis* and *Pseudotropheus* occur. Today a total of about 240 cichlid species are known from Lake Malawi. The actual number of species is probably higher, since new species continue to be discovered.

It should be mentioned that certain species have adapted themselves to extreme water conditions, such as those of Lake Natron and Lake Magadi. Here we find pH values of about 10.5. In these very shallow lakes the water temperature may be over 40° C in various places. It is astounding that in such places fishes such as the cichlid species *Sarotherodon alcalicum* have been observed, which otherwise live in much lower temperatures in other parts of the lake.

Asian Biotope

So far only three cichlid species of one genus are known from the Asian region. One of the two species that have been imported up till now can be reckoned among the dwarf cichlids. It is *Etroplus maculatus*, which is found in southern India and Sri Lanka. Here one finds the fish in rivers and lakes in places with rocky substrates near the coast.

Reed zone on Lake Tanganyika.

Peculiarities

Often, the name cichlid cannot be reconciled with the idea of a peaceful aquarium fish by aquarists or those who wish to become one. One attributes to them universal aggressiveness and pugnacity. To this is added the tendency of the fish to rearrange the aquarium to suit their own tastes. The latter, however, is caused by the fact that the aquarist has done something wrong in the furnishing of the tank. Understandably, most aquarists will not wish to give up

Other fishes, particularly others of the same species, however, are driven from the territory. This occurs particularly intensively during the times of spawning and brood care.

Dwarf cichlids do not have, or have only to a limited degree, the image of being pugnacious and of being bad diggers. To be sure, dwarf cichlids fight and dig like any other cichlids, but these behavior patterns nevertheless do not attract as much notice and also do not cause as much damage to the tank furnishings.

research in their natural habitats is very difficult or often even impossible.

The coexistence of individuals requires mutual understanding. Whether it is a question of territorial formation and defense, of mating, family unity, or schooling, in each case communication is necessary. This occurs through signals of various kinds. The reactions to these signals are innate in the individual fish. Today we are aware of tactile, chemical, optical, and acoustic stimuli that bring about

Fairly large dwarf cichlid aquarium.

the idea of a well-planted, beautiful aquarium when keeping cichlids, although the biotopes of the homelands of the majority of species look quite empty and dreary. All the same, a cichlid tank can be beautiful if one pays attention to a few things.

As a rule, cichlids exhibit a strongly pronounced territorial behavior and create specialized spawning territories for themselves in various ways, including by digging. There also are species-preserving reasons for the so-called pugnacity; one need only call attention to the natural instinct of the fish to defend their territory, their domain.

Therefore, dwarf cichlids will also do quite well in a suitably furnished community tank. Most species can be described as peaceful and they seldom bother other fishes. A number of factors that must be considered are covered in greater detail in the chapter "Keeping." Under optimal keeping conditions, particularly in larger aquaria, dwarf cichlids display virtually the same behavior as in the wild. They have a broad spectrum of different behavior patterns and for this reason are valuable objects for study. We owe the majority of the findings on cichlid behavior to aquarium observations, since analagous

communication in cichlids.

In the aquarium hobby, above all with cichlids, keeping them in pairs is customary. This is not wrong for many species, although other social structures, such as the association of one male and several females or even the association of many individuals of each sex in a school, are also typical for various species. Keeping some species in pairs can be the reason for incompatibility or the failure of spawning attempts. Often the presence of other fishes in the tank is beneficial. Studies have shown, for example, that timid fish lost their shyness after free-swimming fishes, such as tetras, were placed in the tank with them.

Behavior in Food Acquisition

Although it is often claimed that cichlids are predatory fishes, only a very few can actually be considered to be predatory, and dwarf cichlids only in exceptional cases. Even though they will occasionally eat a fry, as is true, after all, of many other fishes, this is not sufficient reason to credit them with a predatory nature. Dwarf cichlids actually eat all sorts of food and, in the manner of

side with it, spit the clump out, and eat the worms individually. Mosquito larvae, especially when the fish are confronted with this food for the first time, are primarily eaten when they leave the water's surface and dive because of a disturbance. But here, too, the fish soon learn the peculiar nature of this food and, if they are hungry, take it from the water's surface. They become acclimated to dry food in the same manner.

In aquaria, males of dwarf cichlids, in the same way as those of larger cichlids, lay claim to the

its fill, continues to guard the food for a while. In all likelihood, this is a distinct aquarium behavior that becomes all the more pronounced the smaller the aquarium is and the more sparsely it is furnished.

Feeding is also well suited for studying the learning ability of these fishes. Thus, one can observe how with time the fish already swim to the usual feeding place when a person approaches the aquarium. Some fish, however, do not come to the feeding place until a particular noise is produced as well, such as

Detail of an aquarium for open spawners.

many other fishes, snap at anything that moves and is of the appropriate size. In the aquarium, the fish have also learned to accept foods that do not move, such as dead and dry foods. In this connection, however, the senses of taste and smell also play a role.

Dwarf cichlids constantly search the bottom for anything edible. If they come across a worm hiding in the substrate, they first pull it out completely and then eat it. If, however, one places a clump of worms in the tank, the fish generally attempt to get the entire clump or a part of it into their mouths. They then swim to the

privilege of being the first to eat their fill before they allow the female to snap at the food. As soon as feeding begins, males of some species immediately chase their females away from the food. The females learn to accept this and try to capture food animals from out of their hiding places. They occasionally dart out a short distance, snap at the food a few times, and then retreat again immediately. One can generally observe this behavior when the fish are fed very sparingly and are always fed at the same place. In addition, it is interesting that the male, when it no longer eats, and thus apparently has already eaten

the lifting of the cover or the uncovering of the aquarium. These fish thus have learned that the mere appearance of the person in front of the aquarium is not necessarily associated with food. One chiefly observes this in cases where one frequently moves in front of the tank. Analogous laboratory experiments, in which interfering factors were purposely eliminated, have clearly confirmed aquarium observations on the learning ability of fishes.

In aquarium keeping, many aquarists do not consider an important point: although on closer observation they can observe repeatedly that the fish pluck at algae, they do not consider feeding them vegetable food. In the biotopes of the homelands of most dwarf cichlids, algae and other vegetable foods are often essential, since at certain times scarcely any other food is available. There are even cichlids and, of course, also dwarf cichlids, that specialize in grazing on the thick dense algae present in their ecotopes. These algae are actually rasped off the substrate by certain species. Other species graze on the expanses of algae in order to feed on the microorganisms living in them. The adaptation to this mode of feeding goes so far that the position of the mouth and the form of the teeth are correspondingly developed.

cichlids, which usually have sufficient space, that is, a necessary territory, at their disposal, it is observed relatively infrequently.

As a rule, fish attract our attention when they fight. We intently follow how the fighting tactics change in the course of the fight and how one of the rivals finally withdraws from the field of combat. Why do the fish actually get in each other's hair in this way? The answer is that fighting serves the preservation of the species: the stronger prevails over the weaker and thereby earns the right to reproduce. This does not seem to make much sense when one finds the body of a rival male in the aquarium, or even that of a female. We should, however, keep in mind that this would scarcely

here is that sufficient food is available for the offspring while they are being cared for. The territory is defended against virtually every intruder, particularly others of the same species that are searching for a mate. What male likes to see his female enticed away from him? So fights take place, which, depending on the strength of the opponents, are only of short duration or which can extend over a longer period of time. The behavior sequence is determined, hence ritualized. Specific signals of one fish call forth specific reflexive answers from the other fish, which in turn act as signals to the first fish and trigger new behavior patterns. These altercations are called ritualized fights in ethology.

Dwarf cichlids react to colors,

Detail of an aquarium for open and cave spawners.

Aggressive Behavior

The aggressive behavior of cichlids is often grounds for describing them as being particularly pugnacious. Although there is hardly a fish species without aggressive behavior, in many groups of fishes it scarcely attracts notice. Because of their size and the need for space often associated with it, the aggressive behavior of larger cichlids is particularly evident. With dwarf

ever happen in the natural habitats, and that it is often the result of incorrect aquarium keeping. In the same way as in the wild, territorial fishes occupy a territory in the aquarium that offers the best potentiality for successful brood care. To this belongs, besides sheltered spawning sites, however, also above all a certain food supply. The size of the required territory is very much dependent on the latter. Important

shapes, and specific movements particularly strongly. We can observe this especially well in the *Apistogramma* species. The initial reaction when a strange member of the same species is sighted is for males to spread all of their fins to their greatest extent and to change their coloration at the same time. *Apistogramma macmasteri* turns a lighter color and a large dark spot simultaneously appears on the

Apistogramma agassizii ♂♂, threat behavior.

Apistogramma cacatuoides ♂♂, parallel posture.

Apistogramma cacatuoides ♂♂, anti-parallel posture.

Apistogramma gibbiceps ♂, threat behavior.

Julidochromis dickfeldi ♂♂, locking jaws.

rear part of the belly area. At the approach of a rival, the anal fin and the rear part of the dorsal fin, as well as the caudal peduncle and the tail fin, are bent slightly upward. Males of *Apistogramma cacatuoides*, *Apistogramma steindachneri*, and several other species additionally spread the gill covers slightly; simultaneously the skin of the throat is pushed out.

In the wild, as well as in the aquarium, if two animals are about the same size and are equally strong, the owner of the territory has the best prospects of driving off the intruder. Often the first phase of the ritualized fight, the threat display, is already enough to cause the rival to retreat. The aggressive behavior patterns that follow this phase are much more frequent in the aquarium than in the wild. Rivals that are not sufficiently intimidated by the threat behavior, and which continue to try to stand their ground in the foreign territory, must reckon with the next phase of aggressive behavior: broadside threatening with tail beating in parallel position to the opponent. At the same time, the rivals spread their fins and abruptly stroke their tails, in which the entire body is brought into use, to cause the greatest possible surge of water to strike the opponent. Each rival's strength is registered indirectly on the other by the magnitude of the water pressure on the organ of the lateral line. With opponents of about equal strength, the impulse to flee in the intruder becomes stronger than the impulse to fight and it will take to its heels.

The fight is continued if neither of the opponents feels inferior to the other. Starting from the parallel position, the rivals circle each other and swim in anti-parallel position; that is, the head is turned toward the tail of the opponent. More tail beating follows in this position. In all probability, the sense organs on the heads of the fish pick up the now stronger water pressure in a different way. Subsequently, both fish swim toward each other with lowered heads, pause for a short time opposite each other, eye to eye, and then abruptly attempt to bite each other's mouths. They now lock jaws, and each of the fish attempts to swim away backwards while pulling the rival along. Jaw locking is carried out with great ferocity. In the process, the fish position themselves horizontally at times, diagonally at others, and shortly thereafter even vertically in the tank. Sometimes they turn around their own axis. As a rule, the decision is reached during this still bloodless phase of fighting. The loser, after it is able to free itself from its rival's hold, leaves the territory with its fins pressed to its body and usually is even chased across the border of the territory.

Because of the limited space in the aquarium, the defeated fish often has no chance to escape. This is not programmed into the behavior inventory of the fish. The only thing that is programmed is that if the other fish does not leave the territory it must want to continue fighting. The victor then generally chases the loser and rams and bites the sides of the body and the fins. The loser also possesses no behavior program for obstructed flight; therefore, it is defenseless against the victor's attacks and continues to be bitten, which ultimately can lead to the death of the defeated fish.

The limited space in the aquarium is also to blame for mate killing with *Nannacara anomala*. Here, however, it is the female that occasionally kills the male through too violent attacks. Identical occurrences have been observed in various *Apistogramma* species. The brood-tending parent—here usually the female—attacks any intruder in the brooding territory, even its own mate. If sufficient space is available in the aquarium, the male has the opportunity to retreat a suitable distance and even to take charge of its duties in brood care. In these species, the male has the task of guarding the outlying areas of the spawning territory, so that the female can devote its complete attention to the care of the clutch.

With territorial neighbors, which stand opposite each other at the border of their territories, one observes a constant back and forth movement along the border. If one of the two has entered the other's territory, the resident fish swims toward the intruder and drives it back across the border, and often even follows it into its own territory. Here, however, the readiness to fight of the followed fish increases again, whereas the follower becomes more likely to take to flight. The fish that has been chased up to this point now turns around and drives out the intruder. At the conclusion of this altercation, in many cases a behavior can be observed that can neither be classified as a willingness to fight nor a willingness to flee. Instead, one observes normal behavior patterns from the sphere of feeding, cleaning, or care. The fishes find themselves in a conflict situation, in which they can neither decide on flight or attack. Such inner tension leads to these so-called displacement reactions, such as suddenly searching for food or clearing away pebbles. They again react to the opponent only when it exhibits renewed aggressive behavior. Comparable reactions can be found in virtually all vertebrates, including man.

Individual males even act aggressively when a female of the same species is introduced. The explanation for this seemingly nonsensical behavior pattern must be sought in territorial defense. The owner of the territory initially tries to scare off all intruders through threat displays. This threat behavior, which would succeed in chasing away a male of the same species, evidently also attracts females. It attempts to weaken the male's aggressive behavior and to demonstrate its readiness to spawn through appeasement displays.

Apistogramma cacatuoides, ♂ courts in front of ♀.

Pelvicachromis taeniatus, *courting pair.*

Nanochromis dimidiatus, ♀ *courts in front of* ♂.

Pseudocrenilabrus philander dispersus, ♂, courtship.

Pelvicachromis subocellatus ♀, normal coloration.

Pelvicachromis subocellatus ♀, courtship coloration.

Reproductive Behavior

In their natural environment, the reproductive behavior of dwarf cichlids depends greatly on seasonal changes in environmental factors. Decisive, among others, are water quality, water temperature, and food supply. In the aquarium, these factors usually are uniformly optimal throughout the entire year.

Reproductive behavior begins with the search for and the defense of the territory. With the female, the development of the eggs and the search for a mate begins. Under natural conditions it can take a long time before the female encounters a male.

If a fish catches sight of a potential mate, which at first is not considered as such but on the basis of the coloration or other signals only as a fish of the same species, hence as a rival, the visual perception immediately triggers the release of specific pituitary hormones, which become evident in the male's threat behavior. The female, attracted by this behavior, now recognizes the potential mate and in its turn weakens the male's initial aggressive behavior. It assumes an appeasement posture in which the fins are pressed to the body. If it is ready to spawn, it stays in the territory, but takes flight if it is not yet ready to spawn. The ready-to-spawn female avoids the male's attacks, which become weaker and weaker and less frequent. The male now courts more and more in its most beautiful colors with spread fins. At the same time, as in aggressive behavior, strokes of the tail fin are dealt out, now, however, with a different goal. In this way the female apparently is signaled that sufficient strength is available to establish and defend a family.

The male's intimidation behavior now hastens the ripening of the eggs in the female. The female begins to look around for a suitable spawning site in the male's territory, while the male now defends the territory more intensively. With open-spawning

cichlids, under certain circumstances the male also takes part in the search for the spawning site.

Ready-to-spawn females of species of *Julidochromis* and *Lamprologus*, to be sure, at first also approach the male with spread fins; however, as soon as they notice aggressive behavior in the male they change their behavior. Trembling intensely, they swim backwards toward the male. These actions are repeated several times in succession. They are a clear signal to the male that a ready-to-spawn female is in its territory. The male's aggressive behavior now is slowly weakened.

Once a female has found a suitable spawning site in the male's territory, it makes the male aware of this in the event that it nears this spot. The male is courted in front of the spawning site. *Apistogramma* females additionally press the fins to the body and deal out tail beats while swimming backwards. *Pelvicachromis* and *Nanochromis* females exhibit a presentation behavior, in which the female curves its body sharply and swims jerkily back and forth while trembling intensely in front of the male. By this means the spawn-filled belly area is displayed particularly prominently.

The female now begins cleaning the future spawning site. With *Apistogramma*, *Pelvicachromis*, *Nanochromis*, *Neolamprologus*, and *Lamprologus* species, this is located, as far as possible, hidden in a cavity. With *Nannacara*,

From top to bottom: The spawning process of typical African mouthbrooders, exemplified by *Haplochromis burtoni*. The male (1) selects a spawning site and "invites" a suitable female. (2) The pair circle and cleanse the selected spawning site. (3) The female begins egg laying. (4) The male goes over the eggs, fertilizing them.

Aequidens, and *Microgeophagus* as well as *Anomalochromis* species, it is found on a rock or a horizontal substrate. *Crenicara* species also use the leaf of an aquatic plant. Mouthbrooders, as a rule, clean the site upon which the eggs will be laid. In some cases they also spawn on rocks, often, however, in a depression. We can thus distinguish cave spawners, open spawners, and mouthbrooders, behavior patterns that are genus-specific.

Cave spawners in most cases find their spawning site under larger rocks lying on the bottom, pieces of wood, or roots. At first it is the female that industriously removes the sand or gravel from under the rock. Each time it fills its mouth full and spits out everything close to the cave. In this way a protective embankment as well as the cave are created. The West African dwarf cichlids are particularly keen diggers, and of them especially *Nanochromis nudiceps* stands out.

Often, it can still take days before spawning takes place. As the time of spawning approaches, the male plays a greater role in the excavation of the cavity and the cleaning of the future spawning site. Shortly before spawning, the female's spawning tube protrudes clearly. Whereas the onset of spawning is not all too conspicuous in *Apistogramma* females, in *Nanochromis*, *Neolamprologus*, and *Pelvicachromis* females one notices it immediately. The females clearly exhibit a

From top to bottom, continued from the facing page: (5) As the female lays the eggs and the male has had an opportunity to fertilize them, she picks them up in her mouth. (6) and (7) The male tantalizes the female by flashing his anal fin so it looks like he is fertilizing more eggs. The female snaps at the egg spots on the male's anal, theoretically getting a mouthful of sperm. (8) The female carries a mouthful of eggs.

particularly full belly area.

Just before the cleaning has been completed, the female glides with its belly over the cleaned site at intervals more and more frequently. One almost gets the impression that it is testing the cleanliness of the substrate with its spawning tube. Without any previous indications, the first eggs are then attached. This occurs on the side wall or the roof of the cavity. During egg laying the male usually waits close by the spawning site. If the female swims to the side, the male approaches the eggs with its belly side and fertilizes them. In addition, one can observe, for example, that fanning movements with the fins are performed by *Apistogramma* species. Subsequently, the fish take turns attaching and fertilizing the eggs. With cave spawners, the clutches of which are of course placed in a relatively secure place, only up to 250 eggs are laid.

With the *Apistogramma* species, in particular, it can be observed that the eggs of a particular pair can have different colors in different spawning acts. This seems to be the result of a change in diet. If the fish are fed predominantly with *Cyclops*, the eggs generally have a bright red color, whereas they can also be white if the fish are fed whiteworms.

After spawning, the female's body looks quite gaunt. It now concerns itself with the care and defense of the clutch. Above all, the eggs are constantly cleaned through sucking. Particularly the females of the *Apistogramma* species have now assumed a yellowish brood-care coloration. The male, on the other hand, as far as possible now no longer stays near the clutch, since it would otherwise be chased away by the female. In small aquaria this can go so far that the male, although it is significantly larger, is bitten to death.

The male again takes charge of the defense of the territory. It is supported by the female in the event that an intruder ventures too

Microgeophagus altispinosa is a typical open ▲
spawner. Here the pair are selecting a spawning site.

The male and female are similarly colored, but the
male has longer unpaired fins. The male is behind the
female. ▲

The male begins scrubbing the selected spawning site
with vigorous attacks with his mouth. ▼

The female helps in cleansing the proposed spawning
site, but her job is also one of close inspection. ▼

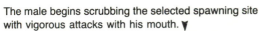

As spawning time comes close, both fish take on a
hue of red. ▼

The eyes of this species are movable and they are
able to see almost straight down, thus assisting
spawning. ▼

Spawning takes place with the female depositing her eggs on the selected spawning site. ▲

In many cases the female lays batch after batch of eggs with the male simply observing. ▲

The male moves closer to the spawning female, seemingly insisting that she leave the site. ▼

As the female leaves, the male begins fertilizing the eggs by swimming over them very carefully. ▼

When spawning is over, the female keeps picking at the eggs, probably removing debris which settles on them. ▼

The eggs polarize within a day or so. A white nucleus develops on one end of the developing egg. ▼

The female continues to hover over the eggs, fanning water over them. ▼

As the eggs develop further, she eats some of them. Those that she eats were probably defective. ▼

A closeup of the developing eggs. ▲

The male does his share, too. ▲

The eggs one day later than the upper photo. ▲

The male also removes defective eggs. ▼

The male fans fresh water over the eggs constantly. ▼

The female searches for a safe place to hide the newly hatched fry. ▲

She prepares a depression in the sand and brings her fry there. ▲

The fry may be moved to a series of depressions until they are free-swimming. ▲

As the fry begin to swim freely, they stay close to the mother (or father). ▲

The free-swimming fry with their mother. ▼

The free-swimming fry with their father. ▼

near the clutch. On the whole, the female is always the more active mate in the direct defense of the clutch.

Depending on water temperature, the larvae hatch 36 to 60 hours after spawning. Often they are husked from the egg membranes by the female at hatching time. I was able to observe this particularly well with *Neolamprologus leleupi*. Larvae of the *Apistogramma* species are brought to a prepared depression. The female will already have excavated several depressions in or outside the cave during the developmental period of the eggs, so that the fry can be moved to a fresh bed frequently.

The fry do not become free-swimming until six to eight days after hatching. They either are very attentively cared for by one parent (*Apistogramma*), both parents (*Pelvicachromis* and *Nanochromis*), or are guarded only very cursorily (*Neolamprologus* and *Julidochromis*). Particularly in species with pronounced brood-care behavior, one can observe how the fry react to specific fin signals of the parents indicating danger: they immediately sink quietly to the bottom.

Open spawners generally seek out spawning sites on flat rocks. In addition, they prefer rocks that are located where plants or plant parts provide cover. In the wild, shallow places in the water with good plant cover often are sought out by several, under certain circumstances, very many, pairs. Males establish their territories here, within which the female, but generally also the male, then searches for a suitable spawning site. If no suitable rock is available, the fish sometimes select the bottom as a spawning site (*Microgeophagus*). They are, however, also fussy with respect to the rock they select. They search for a rock that best corresponds to the color of the eggs. One must then look very closely in order to see the transparent clutch on the rock.

Aequidens curviceps during their spawning. The female and male swim in circles around the spawning site. The female lays eggs fairly continuously while the male fertilizes them after they are laid.

Apistogramma trifasciata × Apistogramma bitaeniata. There are literally hundreds of crosses made between dwarf cichlids. Many *Apistogramma* cross with each other. Most similar mbuna (Malawian rock-haunting species) also freely interbreed even when opposite sexes of their own species are available.

Open spawners, after they have found a suitable spawning site, first excavate several depressions in the substrate. Only later, one to four hours before spawning, do both fish clean the future spawning site. This cleaning is now very intense, however. The cleaning is interrupted by the male's fights in the defense of the territory, in which it is often supported by the female. Now one also clearly recognizes the onset of spawning in the female by the swollen belly area as well as by the protruding spawning tube. Open-spawning females, as well, finally skim over the spawning site with the spawning tube more and more frequently. Then the first, usually clear-as-glass eggs are attached. After that the female swims to the side and the male glides just over the released eggs with its genital papilla, which is now clearly visible, and fertilizes them. Finally, toward the end of spawning, both mates are usually together on the clutch. The female lays more eggs, while the male simultaneously fertilizes the eggs that have already been attached in another place.

A total of up to 700 eggs can be laid. Open spawners lay far more eggs than, for example, cavity spawners, since the clutch of open spawners is exposed to much greater danger. Brood care is also more intensive here. In contrast to some cave spawners, with open spawners both parents care for and defend the clutch. In so doing they continually take turns.

The larvae hatch, depending on water temperature, after 60 to 72 hours and are immediately brought by the parents into one of the prepared depressions. Subsequently, they are often moved to a fresh bed in another depression.

The fry become free-swimming four to five days after hatching. They continue to be cared for by both parents. The parent in charge of the school is usually found in the middle of the school of fry. The fry react very quickly to a parent's

Aequidens dorsiger protecting its swarm of freshly hatched fry. The young have been moved from the original spawning site, once they have hatched, to a depression in the sand.

fin and body movements. The fry react to specific movements of the guarding parent, which signal danger, and all of the fry immediately glide to the bottom where they stay quietly until the all clear signal is given. The coloration of the fry matches the substrate very well. In the same way that both parents tend the school, they also take turns guarding the territory and driving off other fishes that approach too closely. In the event that one

parent cannot handle the intruder, the other gives the fry the sign for danger. The fry sink to the bottom and both parents then fight the opponent. On the other hand, it has also been observed that the guarding parent sometimes swims like a flash out of the school without signaling the fry to sink to the bottom as usual. One can observe this when a food animal is spotted and captured at some distance. As a rule, the fry do not follow the parent on sorties of this kind.

It also happens occasionally that a fry is molested by food animals (*Cyclops*, rotifers). The fry then usually begins to execute abnormal movements. As soon as a parent notices this, it immediately swims over and takes the fry into its mouth. After a few chewing movements the fry is spit out again. After this procedure it swims with the school again with normal swimming movements.

The parents guard their fry so intensely that they will even attack a finger placed near them. Were they not leading fry, they would immediately take flight.

With mouthbrooders, as well, territorial formation and defense by the male is the basis for successful reproduction. In addition, the male generally already establishes the future spawning site at that time. As a spawning site, it chooses either a favorable spot on a partially buried flat rock or digs a depression in the sand. For this purpose it uses either its mouth or its entire body, in that it continually turns in a circle and at the same time throws sand to the outside with strokes of its tail. The male now spends most of its time in the hollow created in this way and defends it and the borders of its territory against all intruders. The size of the defended territory is species-typical and is dependent on territorial conditions.

Whereas any males appearing at the territorial borders are chased off immediately, when a ready-to-spawn female appears

the male is as if transformed. With *Pseudotropheus* and *Melanochromis* species, clear differences in coloration between the sexes exist. The appearance of a ready-to-spawn female, even outside of the territorial borders, triggers an intense courtship behavior in the male. The male begins to court in its most beautiful colors with raised fins. If the female is ready to spawn, it swims slowly in the male's direction. The male then moves toward the spawning depression. Once the female has crossed the territorial borders, the male courts even more intensively. At the same time, it presses the rear part of its body to the substrate and assumes a brighter coloration. Trembling vigorously, it continues to move toward the spawning site. The female follows, of course, but quite hesitently. With this behavior one can speak of a "guide swimming," such as can be observed in, among others, labyrinth fishes.

In species of the genus *Pseudocrenilabrus*, after the female is near the spawning site, the male begins to clean the spawning site in addition to performing a so-called cleaning courtship, in which it presses against the bottom with its hindparts, raises the front of its body slightly, and glides across it while trembling. The female approaches from time to time, but swims away after the male usually begins its frenzied courtship again. Once the female is ready to spawn, however, it also swims to the spawning site and both fish start to circle each other. At the same time, the depression in the sand, if it is not a question of a rocky substrate, is excavated more extensively. If the circling takes place over a rocky substrate, the fish clean the substrate at intervals.

During the circling, both fish—at least it appears that way—attempt to nudge each other in the anal region. The circling becomes more and more abrupt, and then the

Aequidens dorsiger in battle garb. The fry which were nicely disciplined to stay within the confines of the depression made in the sand by their parents (see photo facing page) now start to become free-swimming. They swarm around their parents for an indeterminate period of time. This is the stage at which most of them are eaten by other fishes . . . or even their own parents. For the aquarist, this is the time to remove the parents and raise the fry in their own tank. Many breeders remove the spawning stone with the eggs still attached and hatch the eggs artificially, since parents often eat their own eggs.

female moves forward very slowly and presses its body on the bottom, while the male begins vigorously to nudge the female's belly area. The eggs are released at this time. The female continues to swim in a circle. The male also continues to move, and swims over the eggs and fertilizes them. At the same time, the female now tries to gather up the eggs. Various authors are of the opinion that the eggs are not fertilized until they are in the mouth, namely when the female touches the male's spotted anal fin after taking the eggs into its mouth (egg-spot theory). With these species, fertilization probably already takes place, at least in part, outside of the mouth.

Subsequently, egg laying, fertilization, and uptake of the eggs alternate continually. The eggs generally are of a yellowish color and are relatively large. After spawning, in the course of which up to 100 eggs are laid, the female's throat sac is greatly distended. The female now swims back out of the male's territory, since it otherwise will be attacked constantly, while the male waits for the next lady caller.

Through continual chewing movements, the eggs in the throat sac are repeatedly rearranged. This is of utmost importance for the development of the fry. This absolutely must be considered when the fry are reared artificially. Depending on water temperature, the fry, fully developed and relatively large, are released from the mouth for the first time after 10 to 14 days. At any sign of danger, after a signal from the female, they return to its open mouth.

Pseudotropheus and *Melanochromis* species exhibit almost identical spawning behavior. Males court before the females by swimming jerkily back and forth with spread fins. At the same time, they deal out quite vigorous fin strokes. The surge of water produced causes the female to be jerked back and forth. At intervals the male then positions itself in front of the female with curved body and its entire body starts to tremble. After some time, the male begins with the guide swimming. If a female is not yet ready to spawn, it does not react to this. This is grounds for the male to chase the female away by ramming its body. Ready-to-spawn fish circle each other in the spawning depression, and one can observe the same behavior as in the species of the genus *Pseudocrenilabrus*, except that the eggs are much larger (up to 5 millimeters) and the egg spots in the male's anal fin appear to play an important role in spawning.

Pseudocrenilabrus philander, besides being a very dispersed species in southern Africa, has a very interesting history. Initially the fish was described as *Haplochromis moffati*, in honor of a Reverend Moffat. When the description was changed, the author merely changed *moffati* to *philander*. He neglected to change the sentence saying the fish was named in honor of Reverend Moffat. Everyone knows what a "philanderer" is . . . it may apply aptly to a fish found all over southern Africa, but it could hardly have been appreciated by Rev. Moffat! The fish above is a male. In the photo below, the female begins picking up a clump of about 25 eggs which have been fertilized by the male.

As the female gathers her last egg, her mouth bulges with the spawn which she keeps in her mouth until they hatch as much as a week later (in the cold southern African waters). She further accommodates free-swimming fry too!

Only up to 80 eggs are released; this is, however, a very high number. The fry are let out of the mouth after 20 days at the earliest. At that time they are already about 15 millimeters long and eat small *Cyclops* immediately. Here we are dealing with somewhat more highly developed brood care than with the *Pseudocrenilabrus* species. It is also interesting that the fry usually have the same coloration as the parents from the start.

Mouthbrooding represents the highest form of brood care in cichlids. On account of the protected storage of the eggs and larvae and the already somewhat larger fry, the danger of losses through the most diverse environmental influences is already limited to a great extent. Thus, there is no need to lay a very large number of eggs. In complete contrast to this, the eggs and larvae of open spawners are exposed to dangers of many kinds and are therefore guarded and cared for by both parents. The more specialized brood care is, the less of a bond there is between the parents, and the less time they spend together.

Unusual Behavior Patterns

A quite peculiar behavior also observed in certain dwarf cichlids is the gathering together of *Daphnia* like a school of fry. This has been observed most strikingly with females of *Nannacara anomala*. By this means, excess brood-care instinct apparently is compensated for, which is triggered by the presence of moving objects resembling fry.

The rare matings between fishes of different species of a genus in aquaria would appear to be explainable in the same way. The readiness to spawn of a female, which often is similar in appearance and has the same ritualized behavior, down to a few different details, in the preliminaries to spawning, causes a male, particularly if it has not had a mate for a fairly long time, to accept the offer. Such matings between species are not that uncommon in, among others, the genus *Apistogramma*. Here too it is a question of behavior patterns that can be explained as a displacement reaction to an unfulfilled sex drive. To what extent larvae hatch from the eggs and are then capable of living and reproducing is another matter.

Courting males of *Crenicara*

Crenicara filamentosa is a peculiar fish. The male, the upper fish in this photograph, swims about with his fins clamped and his body swaying in a very unsteady fashion. An experienced aquarist would assume that the fish would be dead shortly! Accompanying this strange behavior is the disappearance of the checkerboard pattern so only a horizontal stripe remains. Believe it or not, this is the sign of imminent spawning. In the photo below, the female is protecting her eggs. She, too, has lost her checkerboard pattern and replaced it with a longitudinal stripe.

filamentosa occasionally exhibit a behavior that leads an observer to assume that the fish will die within a short time. The male swims with fins pressed to its body through the aquatic plants and performs swaying movements, which one recognizes as the unmistakable signs of a fish on death's doorstep. The typical checkerboard pattern has disappeared and the fish displays a dark longitudinal stripe that extends from the mouth as far as the base of the tail. This is, however, the typical sign of breeding condition in the male. The fish now pushes itself over the leaves of the aquatic plants, on which it also occasionally remains stationary (more precisely, lies) as if resting. Since this behavior generally occurs during the time when a ready-to-spawn female appears in the territory or at the border of the

territory, and the male always swims in the direction of the center of the territory, it may be assumed that it is a question here of a type of guide swimming. In this way the female probably is offered a spawning site.

Under aquarium conditions, it can often be observed, in the genus *Aequidens*, for example, that fry that are no longer guarded by the parents form a school time and again. It may be relatively loose, but also so tightly packed that fry stand one on top of the other. This behavior can be classified as defensive behavior against predators. Based on my observations, it always occurs when one enters the room or moves about abruptly just in front of the aquarium glass. Then all of the fry stream together into a swarm, which only begins to break up again when the movement outside of the tank stops.

The coloration of fishes is based on the fact that pigments are embedded in specific cells of the skin. Such cells are called chromatophores. The plasma of these cells always contains only one pigment. According to the pigment in the chromatophores, we distinguish between:

erythrophores - red
iridophores (guanophores) - silvery, iridescent, or reflective
melanophores - black, brown
xanthophores - yellow

Depending on the accumulation of the various one- or multiple-layered chromatophores, the colors are more or less intense and bright. As a result of changes in the angle of incidence of the light, additional variations in color result.

A change in color, which is not uncommon in dwarf cichlids, is under nervous control and is triggered by hormones. The contents of the chomatophores either spread out or clump together. Thus, a dark-colored fish turns a lighter color when the pigments clump together in the center of the cell and turns darker again when the pigments disperse throughout the entire cell.

The most common changes in coloration in dwarf cichlids occur during courtship, spawning, and brood care. The courtship or intimidation coloration, in addition to other specific behavior patterns, is supposed to attract the attention of a mate and to make the male visible even from a distance.

With most females, the coloration of the belly area, in particular, is the decisive color signal for males of their own species. The belly area of ready-to-spawn females often has, in addition to a larger girth, a lighter coloration. In *Microgeophagus ramirezi*, ready-to-spawn females exhibit a red belly area. During spawning and subsequent brood care, females of *Nannacara anomala* assume a conspicuous checkerboard pattern. It could be assumed that in so doing the female is trying lure away

This *Microgeophagus ramirezi* is a man-made color variety called the Golden Ram. This Golden Ram is in the process of spawning. It has always been easier to spawn fishes which were bred in captivity than fishes from the wild.

predators from the clutch as a result of its conspicuous markings. Females of this species are particularly aggressive during brood care. During spawning, the ground color of the females of the majority of *Apistogramma* species clearly changes to a yellowish or, in some species, a bright yellow color. This coloration is supposed to frighten off possible enemies. In most cases, successful brood care is dependent on the female's aggressive behavior. Females of the genus *Aequidens* exhibit a less drastic change in coloration, and often merely become darker. Here, however, both mates take part in brood care.

With the dwarf cichlids of the lakes of East Africa we find almost no changes in coloration. From the time they are fry, the fishes exhibit the coloration they will display throughout their lives, although at first not as intensely.

In contrast to the changes in coloration during brood care, which, as a rule, are retained for a fairly long period of time and which gradually appear, some changes are completed within seconds. Fright coloration is one of these changes. It often resembles the coloration of the fry

and, therefore, is simultaneously a protective coloration. Even upon close examination, fry that lie without moving on the bottom among rocks after an appropriate warning signal from the parents can be detected only with difficulty.

The color changes shown by the majority of species (with the exception of those of the East African lakes) from fry to the fully colored fish are very intense; one need only compare a fry of *Microgeophagus ramirezi* with a fully grown specimen. The larger and more independent the fry becomes, the less it needs protective coloration. Changes in coloration are, however, also still possible in larger specimens. *Apistogramma cacatuoides* males, as long as they have not yet attained their full body size, but nevertheless are already sexually mature, exhibit a quite attractive coloration. With increasing age, virtually all of their bright colors become paler, and the entire body exhibits a gray coloration. In some cases, such age-determined color changes have led to multiple descriptions of species.

Very frequently we encounter changes in the pattern of markings

Microgeophagus altispinosa.

Microgeophagus ramirezi ♂.

Microgeophagus ramirezi ♀.

Microgeophagus ramirezi, golden form.

Microgeophagus ramirezi, yellow-blue form.

Microgeophagus ramirezi, xanthistic form.

with the *Apistogramma* species. One can clearly recognize their mood by their coloration and markings. Here we differentiate not only the gross differences in coloration before spawning and during brood care, but also those of the male in normal, courtship, threat, fright, and appeasement coloration. A few examples are illustrated in the color plates. The different patterns of markings vary depending on the species and could be used, in addition to the traditional methods, to classify the species, if—but this is still a dream of the future—the appropriate documentation is assembled after comprehensive study. Without knowledge of all possible patterns of markings or coloration, misidentifications cannot be ruled out.

Sea Clear Aquariums, as well as other manufacturers, makes small aquariums suitable for keeping dwarf cichlids in a very beautiful setting. (The fishes in this aquarium are marine fishes.) These small aquaria, ranging from one to 10 gallons in size, are available at most petshops. Hidden in the bases of some of them are the heater, pump, etc. It makes for a very tidy setup.

Despite opinions to the contrary, the keeping of dwarf cichlids fundamentally presents no problems. A few conditions are, of course, attached to this assertion. These can be met, however, without great expense.

At the outset, an immediate answer to a question that is asked again and again: may one also keep them in the community tank? Of course! It is beneficial, however, if the dwarf cichlids, which are bottom dwellers, as a rule, do not have to compete with other fishes in the community tank that also require this zone in the aquarium. Dwarf cichlids are a good complement to not-too-large tetras, barbs, labyrinth fishes, and livebearers. One must not, however, place too many fishes in a tank of this kind, no matter how large the tank is! For the genuine aquarist the saying "less is better" should hold true.

In the keeping of dwarf cichlids, one must disregard the formula that is still used in the hobby, according to which the capacity of the aquarium determines the number and size of the fishes. Of course, one cannot completely disregard the size of the fishes, but the size of the aquarium should be selected based only on one principle: the larger the better!

A tank that is as long and deep as possible has proved to be the most suitable for keeping dwarf cichlids. The height is of secondary importance. As far as possible, the tank should not be less than 80 centimeters long. Each aquarist must be governed by the available space in the home as well as possible objections by a spouse. But through perseverance, space for many a fairly large aquarium has been freed up. A long aquarium, for example, fits very nicely in a wall unit and certainly produces a much better impression than knickknacks or souveniers. It must not be forgotten that a larger tank also needs less maintenance, assuming that it is not overpopulated.

Because of their more modern appearance, but also because one can glue a tank together oneself with silicone-rubber cement and the construction is not as time-consuming as that of a metal-framed aquarium, all-glass aquaria have gained acceptance worldwide. This also is reflected in the price. Another advantage is that such aquaria can readily be built in all sizes and shapes, whereby only the thickness of the glass must be paid attention to. Damage to the frame through corrosion is no longer a problem, so that an all-glass aquarium can be used for a much longer period of time. Glass panes can be changed without great expense if they become scratched after some time. The old pane is separated with a razor blade and a new one is glued on. Twenty-four hours later the tank is ready for use again. If the tank is not very high or deep, one fluorescent tube is sufficient as a light source, which, if possible, should extend across the entire tank. Such lighting is adequate for achieving good plant growth in tanks that are not much higher than 35 centimeters. It should go without saying that one covers the tank under the fluorescent tube with an aquarium cover.

The furnishing of an aquarium in which dwarf cichlids will be kept must vary to some extent corresponding to the manner of reproduction and brood care of the particular species. The furnishing of the rear part can be similar in all aquaria. It is best to plant this part with slow-growing aquatic plants (*Anubias*, *Cryptocoryne*, *Echinodorus*, or *Microsorium*). The front part, which in a community tank can simultaneously be the open swimming space for the other fishes, is not planted or is planted only very sparsely with small-growing aquatic plants. Here we can possibly also insert a few resinous bog roots or a halved or whole coconut shell with openings that permit the dwarf cichlids to swim in and out. With whole coconut shells an opening should also be present on top, which is useful for water circulation and which can be smaller than the entrance hole. The opening also allows any air bubbles that accumulate in the shell to escape upwards.

All this applies in particular to the cavity spawners. With the open spawners we do not need the coconut shell. Instead of it, a rock, which is as flat as possible and is located where it is readily visible to us, is placed on the bottom not far from the front glass. The front part of the aquarium can be planted with small-growing plants—such as *Echinodorus tenellus*, *Echinodorus grisebachii*, or *Echinodorus latifolius*—right up to the rock. An *Echinodorus horizontalis*, with its horizontally growing leaves, planted behind the rock can shelter the rock from above. This encourages the fishes to select this place as a spawning site.

For the mouthbrooding dwarf cichlids, one does not necessarily need the rock, since the fishes also spawn in the sand. For the sake of completeness, however, it must be mentioned that particularly with the mouthbrooders from Lake Malawi we do not need to attach as much importance to aquatic plants, but in their place must provide numerous rock structures with sufficient hiding places. In this connection, one must make sure that the rock structures are firmly anchored in the substrate, so that they cannot fall down when the fishes dig. As a substrate for the tanks, one should use gravel with a grain size of from three to five millimeters. It is wrong to mix the substrate with soil, sand, or peat or to place these materials under the gravel and it is also unnecessary for the aquatic plants. In any aquarium containing a number of fishes, virtually all aquatic plants certainly will take root and continue to grow in well-washed gravel.

A filter is necessary for the

furnishing of the aquarium. It should provide as great a flow rate as possible so that a current of water is created in the tank. In addition, the filter's water intake should be located on one side of the aquarium and the return on the other. It is particularly advantageous if one can use a small centrifugal pump to operate the filter. One must make sure that the filter is thoroughly cleaned regularly and at short intervals, since otherwise it will do more harm than good. If a filter is installed in the tank, no aeration is necessary.

If an aquarium is located in a room that has a room temperature of about 24° C even in winter, which, however, may safely fluctuate between 20 and 28° C, it is not necessary to install a heater. In cooler rooms one

Dwarf cichlids are not very fussy about their living quarters. A suitable tank, say 10 gallons, with a light and top glass (to keep in the heat and evaporation, and keep out dust), a few rocks and wood (all of which can be artificial and bought at your local petshop), some plants and a bit of heat.

should use a heater that ensures temperatures of between 23 and 25° C during the day. Temperature fluctuations are better tolerated by the fishes than is a constant temperature, such as one achieves in the tank through the use of a thermostat.

Basically one can use almost any tap water for keeping dwarf cichlids. Even when it is still often stated in the literature that this or that fish needs soft water, let it be said here that this applies to the spawning of certain species, but not for keeping. Furthermore, it is not the water hardness alone that often is decisive for spawning, but rather a low mineral content of the

Hemichromis paynei ♂.

Nannacara anomala ♂.

Nannacara aureocephalus ♂.

Anomalochromis thomasi ♂.

Etroplus maculatus.

Above and below: This schematic shows a setup suitable for cave dwelling dwarf cichlids. **H** = *Hygrophila*. **S** = *Sagittaria*. **I** = *Ceratopteris*. **AU** = *Aponogeton undulatum*.

Above and below: This schematic shows another setup suitable for a cave dwelling dwarf cichlid. **C** = *Cabomba*. **V** = *Vallisneria*. **CB** = *Cryptocoryne becketti*.

Above and below: This schematic shows a suitable layout for an open spawning dwarf cichlid. **L** = *Ludwigia*. **SP** = *Echinodorus intermedius*. **V** = *Vallisneria*.

Above and below: These two drawings are further layouts for a suitable open spawner tank. **V** = *Vallisneria*. **CC** = *Cryptocoryne cordata*. **M** = *Myriophyllum*. **HG** = *Eleocharis*.

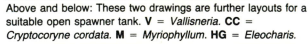

water during this time. Mineral-poor water should not be used for keeping. The *p*H of mineral-poor water is extremely unstable, since the water does not have sufficient buffering capacity. Under these conditions even small amounts of protein decomposition products can have a highly toxic effect on the fishes. It is important, even when using water with a high mineral content, to get used to changing the water regularly. In addition, one should change as much water as possible. No matter how well planted our aquaria are, it is impossible for the aquatic plants to decompose all of the metabolic products that accumulate.

It has proved effective to keep dwarf cichlids in a species tank. Nevertheless, we sometimes deviate from this a bit and place no more than four other fishes in the tank with them. For this purpose it is best to choose species that principally stay at the water's surface or in the upper water levels and which do not grow larger than 5 centimeters.

If only a small tank is available, one will only be able to keep a pair of dwarf cichlids or, with *Apistogramma* species, one male and several females. On the other hand, violent fights between the fishes cannot be avoided, and the death of some specimens is almost unavoidable. If one intends to keep one *Apistogramma* male with several females in a tank, one should make sure that each of the females can establish its own territory. One aids them in this by marking out small territories when furnishing the tank by means of aquatic plants, roots, or rocks. With cavity-spawning species, one will also include a small cavity, such as a coconut shell, in each of these territories.

Besides the principal considerations in the furnishing and stocking of an aquarium, there are also fundamental aspects in the feeding of the fishes. Let it be said beforehand: most errors that are made are not

Petshops carry an amazing variety of dried, frozen and living foods. There is no question that live foods are very important to the health and welfare of dwarf cichlids. While dwarf cichlids might spawn if fed only dried foods, they spawn easier, faster and with a greater percentage of fertility if fed living foods too.
Feed your fish sparingly. It is much better to underfeed than to overfeed. I have NEVER seen aquarium fishes die of underfeeding or malnutrition. I have seen many aquariums become polluted from overfeeding, and all the inhabitants died.

the result of offering the wrong food, but of overfeeding. Unfortunately, even today one can still often find the advice in the literature that one has to feed large amounts of food. This advice is then followed, and the result is obese fishes susceptible to disease, completely disregarding the harm caused by uneaten food that collects on the bottom (growth of fungi, oxygen deprivation). It cannot be repeated too often: feed the fishes as sparingly as possible, and place only as much food in the tank as can be eaten within a short time. One can feed three to four times a day. Dwarf cichlids eat almost anything. *Daphnia*, *Cyclops*, all midge larvae, vinegar flies, *Tubifex*, whiteworms, and even dry food are suitable for feeding. The

above-mentioned live foods can also be offered as frozen food, although one should, as in the feeding of dry food, be particularly careful in determining the amount to be fed.

Even though it is a mark of good breeding to have as little algae as possible in the aquarium, one should not try to control certain species of carpet-forming algae. We will observe that our fishes nibble on the algae very often. In a larger aquarium, the fishes can even feed on the algae and the microorganisms they contain for weeks without additional feeding. It is therefore by no means bad if one occasionally does not feed one's fishes for a few days. The larger the aquarium, the less detrimental is an interruption in feeding.

Some Water Chemistry

Water is not just water; we learn this particularly well when we attempt to spawn species that are considered to be difficult to breed. An example is *Crenicara filamentosa*. In order to be able to judge properly the importance of the spawning water, a few basic concepts will be treated here. In conversations among aquarists, water hardness is a main topic of discussion. The concept of water hardness, however, in itself does not tell us much. What is meant in most cases is namely the total hardness, which above all reflects aspects of the water supply. Soft water has a total hardness of up to 8° dH, medium-hard water 8 to 18° dH, and hard water above 18° dH.

The total hardness (TH) is the sum of the carbonate hardness (CH) and the noncarbonate hardness (NCH). One degree of German hardness (1° dH) corresponds to a concentration of 10 milligrams of CaO (calcium oxide) or 7.14 milligrams of MgO (magnesium oxide) in one liter of water.

The noncarbonate hardness of water, which is also called permanent hardness or sulfate hardness, is based on the concentration of sulfates, chlorides, nitrates, and phospates of calcium and magnesium, whereby calcium sulfate, that is, gypsum, is the principal component. Based on findings to date, the noncarbonate hardness does not have a clear influence on the biological processes in spawning in the aquarium. On the other hand, for aquarium hobby interests, particularly in spawning, the carbonate hardness—the content of calcium and magnesium carbonates or hydrogen carbonates—is of significance. The degree of carbonate hardness is dependent on the amount of carbonic acid present in the water, since this converts the relatively insoluble carbonate into soluble hydrogen carbonate. These chemical compounds can be precipitated through boiling, or even simply by warming the water. The familiar mineral deposits that buiild up in tea kettles are chiefly precipitated calcium carbonate.

For the most diverse reasons, the carbonic acid content is never constant in the aquarium. In a well-planted aquarium, for example, the water generally has a higher carbonic acid content at night than during the day (plants use up oxygen at night and give off carbon dioxide). This can go so far that fishes exhibit symptoms of asphyxiation because of the lack of oxygen, above all towards morning. Many aquarists certainly have been able to observe how their fishes swam at the water's surface in the morning respiring vigorously. It is thus important to aerate well-planted aquaria at night. The carbonate hardness has a negative effect, particularly on the sperm of various fish species—mostly those that are considered to be difficult to spawn. The sperm are weakened and in most cases are no longer able to penetrate the egg. Herein lies the reason why various breeders had unexpected spawning successes even though they had measured a high total hardness: the proportion of carbonate hardness was very low.

How can one keep the carbonate hardness as low as possible? For a long time aquarists softened the tap water with the aid of a cation exchanger. For many fishes, including some dwarf cichlids, the results obtained with this method suffice. The regeneration of the synthetic-resin ion exchanger with salt water is relatively simple. Another method is the dilution of the water with distilled water. In this way the concentraton of minerals is simultaneously reduced. Instead of distilled water, totally desalinated water is generally used today, which one obtains through the use of exchange resins. Many breeders have their own small total desalination installation and prepare their own water for spawning. Totally desalinated water has the quality of distilled water. The regeneration of the exchange resin with hydrochloric acid and sodium hydroxide is, however, expensive.

Whoever wants to spawn certain particularly difficult to breed fishes like the Neon Tetra or *Crenicara filamentosa*, among the dwarf cichlids, in any case needs soft, mineral-poor, hence distilled or totally desalinated water. This leads us to another factor that plays a role in the spawning of fishes: the mineral content of the water. One measures it, as a rule, with an electric conductivity meter, the two electrodes of which are submerged in the water. The resistance of the water to the current flowing from electrode to electrode is called the electrical conductivity and is measured in microsiemens (uS). The conductivity varies as a function of mineral content and water temperature. With precise data, the temperature therefore is also given in a footnote. In practice, these temperature-dependent changes are unimportant for the aquarium hobby. The conductivity value shows us only how high the mineral content of the water is. Which minerals are contained in the water (or not) is also unimportant. In the cases in which we need mineral-poor water, the minerals—of whatever kind—that are still contained in the water may only still be present in concentrations that no longer have an effect on the eggs and sperm. One will always be on the safe side if one adds one cup of tap water to each 10 liters of totally desalinated water, although one also finds water in the wild which is comparable to totally desalinated water. One can then dispense with the measurement of the electrical conductivity.

Fishes, of whatever species, may not be kept in mineral-poor water of this kind for a fairly long time. It has far too little buffering capacity and often reacts relatively rapidly to the various influences on water chemistry. Thus, the *p*H can quickly change to a value that is dangerous for the fishes and can result in their loss. The fishes, therefore, are kept in water of this kind only briefly at spawning time.

The acidity or alkalinity of water is given in *p*H. Water that reacts neutrally has a *p*H value of 7. A *p*H of less than 7 means acidic water, and greater than 7 alkaline water. In the aquarium hobby we deal with values of between 4.5 and 8. For the spawning of dwarf cichlids in general a neutral *p*H of about 7 is adequate. The measurement should at least be made with the Czensny indicator. The most precise measurements are obtained by electrical means. Because of the high purchase price, such devices are seldom used in the aquarium hobby.

The *p*H value is subject to natural fluctuations as a result of the most diverse influences in the aquarium. The aquarist should not try to influence the *p*H. Although, aside from the water hardness, the *p*H is discussed more often than any other aspect of water chemistry in the literature, because of the numerous successes at a variety of values it is to be doubted whether better results could be obtained through the artificial shifting of the *p*H than at neutral values. If one wishes to experiment, any acidification or increase in the alkalinity of the water should be carried out with the utmost care. Extreme caution is called for, particularly when we have water with minimal buffering capacity. In no case should fishes be in the tank during these manipulations. After a period of time one should measure the *p*H once more before adding the fishes.

The ammonium content of the water is best regulated through partial water changes. Ammonium compounds, which are created through protein decomposition (excrement from the fishes, bacterial decomposition of food remains), are harmless in the acidic to neutral range but are converted into toxic ammonia at

Your local petshop will almost certainly have a suitable selection of testing kits and chemicals to treat the common aquarium water problems.

alkaline *p*H values. When performing partial water changes one should make sure that the fresh water has a neutral or slightly acidic *p*H. A measurement of the nitrate, nitrite, or ammonium content is expensive and is unnecessary if frequent water changes are carried out. In well-planted tanks, the plants take up a significant portion of the nitrogen compounds as nutriments.

Species for Which no Special Requirements are Necessary for Spawning

If we follow the instructions for keeping given in the previous section, we will, as a rule, already have laid the foundation for successful breeding. On principle, however, it should also be mentioned that spawning on command almost never succeeds. Cichlids that we assign for spawning always need some time to become acclimated to each other, on the one hand, and to the new living space, on the other hand. Of course, cases are also known in which the fishes spawned shortly after they were put in the spawning tank. In these cases various conditions usually were favorable, but more often the fishes simply were ready to spawn at that time.

Spawning tanks can be, without further ceremony, the furnished aquaria in which we keep the fishes for a fairly long period of time. One can, however, also use special spawning tanks in which no furnishings are found besides the necessary spawning substrate. A filter or a discharger belongs to the essential technical equipment.

Some species do not need any changes in the tank to spawn successfully. To them, among others, belong:

Aequidens curviceps
Aequidens dorsiger
Anomalochromis thomasi
Apistogramma borellii
Apistogramma reitzigi
Julidochromis dickfeldi
Julidochromis marlieri
Julidochromis ornatus
Julidochromis regani
Julidochromis transcriptus
Melanochromis exasperatus
Neolamprologus elongatus
Neolamprologus leleupi
Nannacara anomala
Pseudocrenilabrus multicolor
Pseudocrenilabrus philander
Pseudotropheus aurora
Pseudotropheus elegans
Pseudotropheus lanisticola
Telmatochromis bifrenatus

Species that Place Special Demands on the Water for Spawning

This second group is made up of the species that need soft water with a low carbonate hardness and a low mineral content for spawning or that are spawned most successfully in water of this kind. To them belong, among others, the following species:

Apistogramma agassizii
Apistogramma gibbiceps
Apistogramma bitaeniata
Apistogramma sweglesi
Apistogramma trifasciata
Apistogrammoides pucallpaensis
Nanochromis dimidiatus
Nanochromis nudiceps
Nanochromis robertsi
Pelvicachromis roloffi
Pelvicachromis taeniatus
Pelvicachromis subocellatus
Taeniacara candidi

Here too, on principle the tank used for keeping can serve as the spawning tank. The water should have a carbonate hardness of up to 6° dH and a conductivity of not more than 400 *u* S. Such water is best obtained by mixing tap water, which in most cases is medium-hard to hard, with totally desalinated or distilled water. As a rule, one cannot go wrong if one adds 1 liter of tap water to 8 liters of totally desalinated water. The water that flows out of faucets in Europe is, to be sure, extremely variable in hardness, but almost everywhere its mineral content is very high in comparison to the natural bodies of water of the homeland regions of dwarf cichlids. For this reason, it does absolutely no harm if we achieve a very low water hardness; the mineral content of the water in all cases will be favorably influenced in this way.

With very hard water a different ratio of components must be selected so that one can still reach the required low hardness. For that purpose, however, it is necessary to know the hardness of the tap water. After measuring it, we can calculate the ratio of components we must choose by means of the component cross, as is shown in the following example:

Our tap water had a total hardness of 40° dH and a carbonate hardness of 15° dH. Totally desalinated water has a hardness of 0° dH. Although we know that the total hardness is not as critical as the carbonate hardness, we first calculate the ratio of components for water with a desired total hardness of 10° dH. We draw the following diagram:

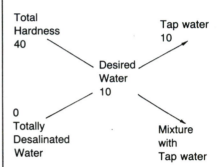

In each case we draw an arrow from the degrees of hardness of the starting water through the hardness value of the desired water, subtract the desired hardness from the total hardness of the tap water, and obtain the necessary proportion of totally desalinated water, in our case 30 parts. Since totally desalinated water has no hardness, nothing is subtracted from the value of the desired water. In this way we arrive at the second value, the proportion of tap water, here 10 parts.

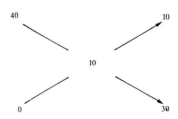

Therefore, we must mix 10 parts tap water with 30 parts totally desalinated water to obtain water with a hardness of 10° dH.

If we wish to have water with a carbonate hardness of 4° dH, we obtain the following diagram:

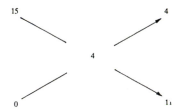

We mix 4 parts tap water with 11 parts totally desalinated water to obtain water with a carbonate hardness of 4° dH.

We keep the fishes in tap water until we notice that they have started their spawning preparations, or that we clearly recognize that the female is filling up with spawn. Only now do we perform an almost complete water change. We lower the water level so far that the fishes are barely still able to swim properly. One then allows fresh water of the indicated spawning quality to run slowly into the tank, through a thin hose, for example. It is beneficial to insert the end of the hose in the gravel. When the water level is so high that the filter functions again, one lays the end of the hose on the filter. Excess air, which is still in the water and which could possibly be discharged onto the fishes, will for the most part be eliminated in this way.

In this manner the fishes will scarcely be disturbed and will also slowly become acclimated to the new water. It is like a heavy shower of rain for them and stimulates the readiness to spawn. From personal experience, I can state that even a faster water change does no harm, but one should go about it carefully.

This water change often stimulates a pair to begin spawning after a short time. Depending on how accurately we estimated the time of spawning before the water change, we sometimes need only wait hours, at other times, however, also days, for spawning to begin. We continue to feed the fishes an adequate and as varied as possible diet. The feeding of whiteworms has proved to be beneficial during this time. Should spawning happen to be delayed

One of the color varieties of *Pelvicachromis taeniatus*, a female, guarding her eggs in her coconut shell cave.

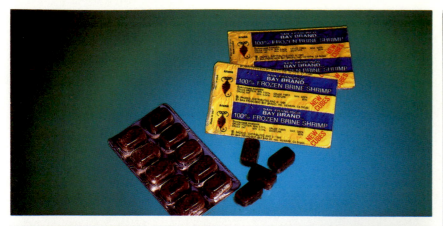

Brine shrimp, *Artemia* species, are a fish breeder's best friend. Without *Artemia* breeders would be severely hampered. The freshly hatched nauplii are the best first food for many fishes, including but not limited to dwarf cichlids. The adults make excellent food for larger dwarf cichlids . . . and intermediate sizes are suitable for growing dwarf cichlids. They are also safer than other live foods since they originate in very salty lakes which do not harbor any freshwater parasites; they are also hatched in brine, thus again being safe from harmful freshwater organisms. Most petshops carry brine shrimp eggs in many forms and sizes; they might also have the live or frozen forms, as well as the freeze-dried forms. Brine shrimp are ubiquitous, being found all over the world.

somewhat longer, however, it is appropriate after at most a week to replace the water, at least partially, with fresh spawning water.

Once the fishes have spawned, we wait until the larvae hatch and then change the water again. Now, however, we use tap water.

After the fishes spawn, one should also make sure, above all with cavity spawners (*Apistogramma*) in relatively small aquaria, to remove the male. This is unnecessary in large aquaria with sufficient hiding places.

If one has fishes that do not care for the clutch properly or that even eat the eggs, in subsequent spawning attempts it is best to remove the clutch from the tank after spawning and to transfer it to a rearing tank. In the rearing tank one should use water of the same quality as in the spawning tank. It is best to position a filter discharge right next to the clutch to provide for continuous water movement. After the larvae hatch, one removes the spawning substrate

from the rearing tank. It often happens that all of the larvae do not come out of the egg membranes properly. By swishing the spawning substrate back and forth in the water one can help the fry to hatch. The fry become free-swimming about four days after hatching. At that time they must be fed with the finest newly-hatched *Cyclops* or *Artemia salina* (brine shrimp).

For the spawning of certain dwarf cichlids, for example *Crenicara filamentosa*, it is necessary to use water with, if possible, a carbonate hardness of 0° dH and a very low electrical conductivity of less than 200 u S. One can proceed here in exactly the same way as with the previously mentioned group, but with one qualification: a tank without substrate and without aquatic plants should be used. It has also proved beneficial to begin changing the water somewhat earlier and gradually to match the future spawning water through the partial addition of

totally desalinated water. One starts with this as soon as the female begins to fill up with spawn. At first half of the water, and a few days later almost all of the water is replaced with totally desalinated water. To 10 liters of totally desalinated water one adds only about one cup of tap water. It has also proved beneficial to filter the spawning water through peat. Then the eggs do not become attacked by fungus as quickly.

During the time before spawning the fishes are best fed with whiteworms. All other food animals die very rapidly in this spawning water and only foul the water. Of course, one can also feed with *Cyclops*, but then one must really count out how many are given so that all of them are eaten.

Many of these species also spawn in water that has not been pretreated. The yield from such spawning attempts is, however, often low.

Aequidens curviceps.

Aequidens dorsiger.

Aequidens spec.

Aequidens spec.

Crenicara filamentosa, ♀ on clutch.

Crenicara filamentosa ♂.

With today's offering of exotic fishes it is striking that, besides the wild forms, more and more fishes are offered that differ, sometimes greatly, from these in form and coloration. The close kinship to them is, however, still readily apparent. These are fishes possessing either greatly elongated fins (veil forms), a different coloration—usually more striking than that of the normal form, or which even differ in build. There is a tendency to speak of strains with all of these variations, but only in few cases are they actually the result of systematic work by breeders. Much more often we are dealing with accidental mutations that later are used for crosses.

In heated discussions among aquarists, up until a few years ago the question was argued whether or not it is sensible to reproduce every mutation or hybrid that appeared, thereby increasing the offering of exotic fish forms. Today, under the pressure of continually stricter protective regulations, which one must also fully support in principle, it is necessary to worry about how one will be able to satisfy the demand for new forms of exotic fishes without the importation of new species. The spawning of the still available wild forms remains, as before, of decisive importance for exotic fish breeding; however, the assortment cannot be increased in this way. Today there is agreement that the further reproduction of mutations, hybrids, and strains also represents a very good possiblity for increasing the assortment. Whether these fish forms are beautiful or ugly will always remain a question of taste, but, as everyone knows, one should not argue over taste.

Although there are still relatively few known mutations, hybrids, and strains of dwarf cichlids, this certainly does not mean that there is less of a potential to that end. Who would have thought only a few decades ago that there could ever be such a wide assortment of strains of the angelfish as are available today. Among dwarf cichlids, as well, attractive spawning results have already been achieved.

What appears to be so simple in the aquarium, at least in theory, as the breeding of specific color mutations or selective breeding, is hardly to be expected under natural conditions. Here the environmental conditions determine whether or not a newly created form is viable. Environmental conditions, as a rule, are also extremely severe for fishes, and hence only those fishes that are optimally adapted to the ecological conditions can hold their own. Any variation that prevents the fish from performing the behavior characteristic of the species or even only hinders it, such as veil fins, substantially reduces the chance of survival.

With mutations, hybrids, and strains, in most cases we think only of the externally visible changes in the fishes, all the more so because these, of course, are in the forefront of the aquarium hobby. Often the visible changes are accompanied by invisible ones in the fish's body or by changes in behavior that are generally overlooked in the aquarium hobby.

Mutations

In the aquarium hobby, here and there differing body characters are found in fishes. For example, entire broods of offspring can exhibit deficiencies or malformations. This is often the result of incorrect aquarium keeping. Poor diet and lack of water changes during rearing could be the cause. If the poor aquarium conditions are maintained, such stunted specimens can continue to produce offspring over several generations in which these deficient characteristics are retained, until completely normal fishes develop again if optimal conditions are provided. This change in the phenotype, which is called a modification, is influenced by external factors and is not inheritable.

Random inheritable changes, which do not appear as the result of crosses, are called mutations. They occur either spontaneously or are induced through the action of mutagenic factors (for example, X-rays, colchicine, hydroxylamine). Through mutations, build and coloration, size, metabolism, and even behavior patterns can be changed. Inheritable changes in behavior are not as easy to recognize as changes in coloration or even albinism. In the aquarium, the predatory factor is lacking and hence the survival of such fishes is dependent on the aquarist alone.

In practice, mutations are usually crossed with the original form in the aquarium. In this way it is subsequently attempted to stabilize further positively assessed changes through selective breeding.

The most familiar mutation of dwarf cichlids could well be the gold coloration of *Microgeophagus ramirezi*. Also known are the xanthic (yellow) varieties of *Cichlasoma nigrofasciata* and *Etroplus maculatus*.

Strains

It is by no means rare that one finds fishes that exhibit fairly small mutational changes among a large number of offspring. These can be, for example, slight variations in coloration. Depending on whether you are of the opinion that it is or is not a question of a character that improves the appearance, such a fish is either used for further breeding or is excluded from it. If attractive additional or varying characters are found, the thought occurs to breed this character into the fish even more clearly. One therefore sets a breeding goal and through selective breeding (selection of the breeding animals) from generation to generation achieves an improvement in the desired character, unless this character is

Apistogramma bitaeniata ♂.

Apistogramma borellii, wild-caught specimen, color variety.

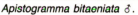

Apistogramma borellii, wild-caught specimens, color variety.

Apistogramma borellii, wild-caught specimen, color variety.

Apistogramma cacatuoides, wild form.

Apistogramma cacatuoides, strain.

Apistogramma cacatuoides, strain.

Apistogramma cacatuoides, strain.

Apistogramma cacatuoides, courting.

genetically coupled with a lethal factor or the fishes in some other way become incapable of living through the strengthening of the character. Various particularly beautiful color varieties of *Apistogramma cacatuoides* and *Apistogramma agassizii*, which were produced in this way, are pictured in the color plates. These are not new species or subspecies; through selective breeding an already present character was merely positively reinforced according to human points of view.

One can scarcely imagine how much work is involved in breeding these fishes in the most diverse forms and in the genetically purest form possible. Generations have worked on many of the forms available today. This is also grounds for warning against aimless genetic experiments with animals. The preservation of a wild species in the aquarium is at least as commendable an achievement as the breeding of a new color variety.

Hybrids

Most of the strains offered on the market are the result of crosses between mutants. Long-finned, xanthistic, and albino fishes always occur only sporadically as isolated mutations. The resulting character must be maintained through crossing with the normal form. If it is not dominant, it usually disappears again from the stock. If the new character has generated interest with aquarists, the thought occurs to carry over this character to other species. As long as it is a question of a genetically very closely related species, this cannot be ruled out. Thus, one already finds today long-finned forms of many species (not every veil form need necessarily be the product of crosses). Breeders have been able to achieve considerable success with the live-bearing tooth-carps of the genus *Xiphophorus*, the genetic variability of which is very great.

Although very little has been done up till now in the area of hybrids with dwarf cichlids, this is at least in part because they have not been bred so far in the same numbers as live-bearing tooth-carps. The coming years will show that here, too, hybrid forms that eclipse their original forms will be possible. The increased importation of color varieties of various *Apistogramma* and *Pelvicachromis* species will be a good foundation for this.

An astounding cross already succeeded years ago in the Soviet Union. *Julidochromis marlieri* was crossed with *Neolamprologus elongatus*. Such hybrids apparently can only be achieved through special manipulations. They are not the goal of hybridization from the point of view of the aquarium hobby, especially since offspring from hybrids of different species or even genera are sterile, as a rule.

Systematic breeding requires comprehensive knowledge of genetics, which in increasing measure is assumming a key position in biology as a whole.

It is important, however, that genetically pure wild forms are preserved as a genetic reserve and are not only offered as hybrid products, which is already the case with certain species.

Apistogramma agassizi male.

Apistogramma agassizii, normal coloration.

Apistogramma agassizii, strain.

Apistogramma agassizii, strain.

Apistogramma agassizii, strain.

Apistogramma agassizii ♀, mutation with elongated tail fin.

One day one will stand in front of one's aquarium and observe how some fishes act strangely. A fish rubs abruptly sideways over the substrate or the edge of a rock and finally stops in the tank, its body slowly undulating. Upon closer inspection one sees white deposits or small dots on the fins and skin, tattered fins, noticeable difficulty in respiration, or other symptoms of a fish disease. One is usually very surprised by such observations and is not sure what one should call the particular fish disease or how to control it. If something is not done immediately, however, with certain diseases the entire fish population of the tank can perish. The earlier one recognizes and treats diseased fishes, the greater are the chances of a cure, although in the aquarium hobby it sometimes is not worthwhile and often is impossible to save the fish. The aquarist, as a rule, cannot make a diagnosis, but rather can only assume what sort of illness he is dealing with. Therefore, with the frequent occurrence of the same symptoms it is advisable to bring one or several afflicted fishes to a fish-disease specialist while they are still alive. He will make an accurate diagnosis and recommend what should be done.

Prevention should be in the forefront for the aquarist. It is the best method for avoiding illness in fishes. Fishes as such possess a good system of resistance. Often the inadequate conditions in the aquarium are to blame for the outbreak of various fish diseases. Medications should always be an emergency solution since they not only cure, but often can do harm as well, above all with imprecise dosage. Frequent water changes, varied and not too generous feeding, a sparse population in the tank, as well as keeping dwarf cichlids together with other suitable fish species are the best preventive measures. Diseases are caused by infections of various kinds. In general, therefore, one should avoid injuries and damage to the skin and maintain the stability of the mucous membrane barrier through favorable water conditions (free of harmful substances, rich in oxygen). Through regular, close observation the aquarist quickly will be able to recognize an infection on the basis of specific symptoms.

The following pointers, which have proved their worth in practice, are meant for diseases that can be recognized externally and that can be approximately classified.

The following diseases are among those most frequently encountered with dwarf cichlids:

Infectious dropsy turns up now and then, particularly with *Apistogramma*, *Microgeophagus*, and *Pelvicachromis* species. The initial symptom is the swollen belly area, which in the course of time becomes more and more swollen. In its final stage, dropsy is usually accompanied by protruding scales and somewhat bulging eyes. Treatment would be possible with antibiotics, except that the illness, by the time one notices it, is already so far advanced that there no longer is any chance for a cure. Therefore, it is recommended to remove and destroy fishes that exhibit symptoms of dropsy.

The appearance of ulcers has various causes. They can be considered as secondary conditions of infectious dropsy or as a symptom of existent tuberculosis. The distinct thickening of an area of skin is the first symptom of this illness. Later a white spot appears on the thickened area of skin, which becomes larger and larger and finally separates from the skin. As a rule, an inflammation appears in this place, whereby the red muscle also is exposed. This disease is almost impossible to cure in the aquarium. Thus, in this case as well, destroying the fish is the only thing that can be done, all the more so because other fishes can be infected.

The same applies to fishes with pop-eye (severely protruding eyes). Pop-eye often appears in conjunction with dropsy or tuberculosis. The *Pelvicachromis* species are often stricken with it.

Occasionally, fishes become infected with one-celled organisms that produce a filmy cloudiness (*Costia*) or small white dots (*Ichthyophthirius*) on the skin and fins. Fishes infected with *Costia*, as a rule, exhibit loss of appetite, clamped fins, swaying movements while swimming, and rub against the substrate or objects. With a heavy infestation, a whitish-bluish film is visible on the skin that consists of dead skin cells and parasites. Treatment, which should take place in a separate, unfurnished tank, in many cases is successful only if the fishes have not yet become too greatly weakened. A half-hour bath in a table-salt solution has proved effective. For this one uses about 10 grams of table salt for each liter of water. Treatment with a malachite-green preparation has also shown good results. Also to be recommended is an acriflavin bath (0.1 to 1.0 grams per 100 liters of water for two to three days) or raising the water temperature to 30° C until the cloudiness of the skin disappears.

Ichthyophthirius infestation is also called white spot disease since it is easily recognized by the white dots on the fins and skin. Additional symptoms include difficulty in respiration and clamped fins. Treatment usually only makes sense when the fishes exhibit only a few small white dots. It must take place immediately. Since the reproductive cycle of this skin parasite is completed outside of the fish, it is possible to find the free-swimming parasites in large numbers in the aquarium water. Therefore, the first measure for preventing reinfection is a radical-as-possible water change or the transferring of all of the fishes to a parasite-free and strongly filtered tank. After that the treatment with chemotherapeutic drugs follows, whereby malachite-green preparations have proved to be quite effective. Since the parasites are located under the fish's mucous membrane, they are not killed there by this preparation. They are not destroyed until, after further phases in development, they search for a new host as free-swimming parasites in the open water.

If one wishes to avoid chemotherapeutic measures, one can also use the following extremely time-consuming procedure. One must have several strongly-filtered tanks at one's disposal. The fishes are transferred to a different tank every 8 to 12 hours. In this way the free-swimming parasites have little opportunity to infect the fishes, and so die out. One should transfer the fishes regularly for at least a week and make sure that no additional spots appear on the fishes for a period of several days before discontinuing treatment.

There are also protozoans, such as *Octomitus*, which are parasites of the blood or which occur in the intestine, gall bladder, and other organs. An *Octomitus* infection, which is not rare with Malawi and Tanganyika cichlids, can be recognized by dark coloration, emaciation, and slimy, whitish excrement. Acriflavin is suitable for the control of these parasites (0.1 to 1.0 grams per 100 liters for two to three days).

The abrupt rubbing of the gills against objects can indicate an infestation of the gills with protozoans or flukes. In the course of the illness, loss of appetite and emaciation are typical symptoms. Treatment is as with a *Costia* infestation or with Trichlorphon.

An illness that almost always appears only with wild-caught fishes and that one at first does not regard as an illness, is the infestation with nematodes. Only after some months can one observe the at first just barely 1-millimeter-long nematodes hanging out of the anus. With increasing growth, these can extend up to 10 millimeters

Petshops carry complete centers for the identification and treatment of the most common aquarium fish diseases. If the common drugs and chemicals available are not successful, you are best advised to dispose of the fish since they will probably die anyway and might even infect other fishes in your care.

outside of the body. It is generally a question of *Camallanus* worms when we observe such an infestation. Despite the use of the most diverse measures, I was not able to control them. One should therefore remove and destroy infected fishes at the first sign of infestation. The following symptoms apply to a nematode infestation: emaciation with complete loss of appetite, rubbing, and slimy, whitish excrement. Whereas one can already suspect a *Camallanus* infection by the worms hanging out of the anus, other nematodes can be detected only through the microscopic examination of the excrement or the intestinal contents. Treatment is possible using food animals soaked in Niclosamide.

In closing, a comment on the possible effect of medications on wild-caught fishes. In normal cases, the fishes reach us by way of the collector through collecting stations, exporters, importers, wholesalers, and dealers. Today fishes often are already preventively treated with the most diverse medications at the collecting stations. In so doing, not infrequently they are given an overdose. At every additional stop on their way to our aquaria a new treatment takes place. The sex organs of such fishes are sometimes damaged to such a degree that they cannot spawn even under optimal aquarium conditions. Based on my own experiences, right into the 1970s, when this preventive treatment with medications did not yet take place, the spawning of wild-caught fishes often was easier than that of aquarium-bred fishes. For that reason, breeders, in particular, insofar as there is a possibility of curing diseased fishes, should treat them with natural measures (water changing, raising the temperature, repeated transferring to other tanks, adding salt).

Pop-eye in *Pelvicachromis pulcher*.

Ulcer formation in *Apistogramma nijsseni*.

Costia infestation in *Aequidens dorsiger*.

Microgeophagus ramirezi with dropsy.

Apistogramma cacatuoides with nematode infestation.

Diseases of dwarf cichlids.

Description of the Species

Melanochromis johanni.

Genus *Aequidens* Eigenmann and Bray, 1894

Explanation of the scientific name: Refers to the dentition of the species of this genus—with teeth of equal length.
aequus (L.) = even; *dens* (L.) = tooth.
Original description: Eigenmann, C. H., and Bray, W. L. (1894): A revision of the American Cichlidae. *Ann. New York Acad.,* 7, p. 616.

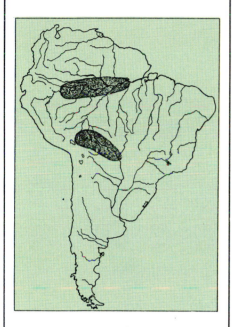

Distribution: Northern South America and Central America.
Number of species: So far about 35 species are known.
Generic characters: The majority of the species of this genus were formerly classified in the genus *Acara* Heckel, 1840. In contrast to the similar-looking species of the genus *Cichlasoma*, they possess only three spines in the anal fin. The lobe on the first gill arch is lacking; no gill rakers are present. The fishes have conical teeth of virtually equal length.
Total length: 80 to 300 millimeters.
Type species: *Aequidens tetramerus* (*Heros tetramerus* Heckel, 1840).

Aequidens curviceps (Ahl, 1924) Flag Cichlid

Explanation of the scientific name: Refers to the curved head profile (forehead).
curvus (L.) = rounded, curved, arched; *kephale* (Gk.) = head.
Other names in the literature: *Acara thayeri* Arnold, 1911
Acara curviceps Arnold, 1924.
Original description: Arnold, E. (1924): *Mitt. Zool. Mus. Berlin,* vol. 11, p. 44 (*Acara curviceps*).
Distribution: Virtually the entire Amazonas river basin.
Habitat: Sheltered sites with fairly small, often fast-flowing bodies of water with overhanging riparian growth or exposed roots.
First importation into Germany: A single specimen in 1909 by Siggelkow (Hamburg); several specimens in 1910.
Characters:
fins: D XV/7, A III/7, P 15.
scales: mLR 23–24.
total length: male up to 80 millimeters; female up to 60 millimeters.

Aquarium keeping: Flag Cichlids are best kept in pairs in community or species tanks. A species tank should have a length of at least 70 centimeters and a capacity of 60 liters. No particular demands with respect to water hardness. Water temperature between 24 and 28° C. Background and sides of tank should be thickly planted. In the foreground a flat rock is placed under fairly large plant leaves or roots. Pairs kept separately sometimes display aggressive behavior toward the mate. One can take action against this by keeping them with several other fishes that prefer to stay at the water's surface or in the middle water levels. In this way a buildup in aggression can be avoided, since the fish can continually let off steam.
Spawning: The fish spawn on a flat rock that they have cleaned intensively beforehand. The clutch can consist of up to 300 eggs. The

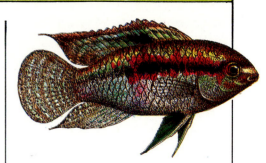

eggs are translucent amber-colored. Both parents alternately tend the eggs, whereby the female, however, cares for the eggs more intensively and the male principally looks after the surroundings of the spawning site, that is, primarily takes charge of territorial defense. During this time the female excavates small, shallow depressions close to the spawning site. The larvae hatch about two days after spawning, or, in most cases, are sucked from the egg membranes and placed in the prepared hollows by the female. Sometimes it can take over an hour before the last larvae have hatched from the egg membranes. Once all of the larvae have been placed in the depression, they sometimes are immediately moved to a different hollow. This takes place very often in subsequent days. Each time the larvae are moved to a fresh bed, one of the parents—usually the female—positions itself over the hollow and fans fresh water onto the larvae. Four to five days after the larvae hatch the fry become free-swimming. The parents, which now assume a somewhat darker coloration, guard the school of fry very devotedly. At any given time one of the parents in turn swims in the middle of the school and leads the fry, while the other devotes its attention to territorial defense. One feeds the fry with newly hatched *Artemia* or *Cyclops*. Each evening the fry are brought back to one of the hollows by the parents and are fanned by the female, which settles down right next to the hollow during the nightly rest period.
Comments: On account of the

Aequidens curviceps spawning. Note the bulging ovipositor of the female as she deposits her spawn. The male stands by ready to fertilize the eggs after the female has laid them.

very extensive range of this species, specimens from different collecting areas exhibit variable coloration.

Aequidens dorsiger
(Heckel, 1840)
Red-breasted Cichlid

Explanation of the scientific name: The specific name apparently refers to the markings of the dorsal fin; there is no explanation by Heckel (1840).

Other names in the literature:
Acara dorsiger Heckel, 1840
Acara dorsigera?
Parvacara dorsiger Whitley, 1951
Aequidens dorsigerus?

Original description: Heckel, J. J. (1840): Johann Natterer's neue Flussfische Brasiliens's nach den Beobachtungen und Mitteilungen des Entdeckers bescrieben. Ann. Wiener Mus., vol. 2, p. 348–350 (*Acara dorsiger*).

Distribution: After Heckel (1840), the swamps around Villa-Maria on the Rio Paraguai. According to the accounts of various collectors, the range is apparently quite large (western and northwestern Bolivia as far as Brazil). The author collected the species in a small tributary of the Lago Mandiore, about 100 kilometers north of Corumba in the state of Mato Grosso (Brazil).

Habitat: Current-free coves of small rivers as well as in standing residual bodies of water without

Aequidens dorsiger preparing to spawn. The male, as in most *Aequidens*, has longer unpaired fins than the female.

aquatic plant growth, but with overhanging vegetation or branches in the water. Water depth between 30 and 100 centimeters. Clear, amber-colored water, virtually without hardness, with *p*H values of between 5.5 and 6.5.

First importation into Germany: 1977.

Characters:
fins: D XV/9, A III/8.
scales: mLR 24
total length: male up to 80 millimeters; female up to 60 millimeters.

Sex differences: Male somewhat more elongated than the female, dorsal and anal fins are more extended in older males, and the head form is more blunt. Females produce a more compact effect, the belly area being rounder and the head being more pointed than in the male. Younger fish are difficult to differentiate.

Aquarium keeping: The same conditions should be provided as with *Aequidens curviceps*. Supplemental feeding with vegetable food (frozen spinach) or dry food with a high ratio of vegetable components is necessary. Planting the tank with tender-leaved aquatic plants is not recommended as the fish will chew on them.

Spawning: Without problems, as with *Aequidens curviceps*. Fish that are ready to spawn prominently display the red coloration of the breast and belly areas typical of these fish. During spawning, particularly in the female, virtually the entire ventral part of the body turns a dark color. Up to 1000 clear-as-glass eggs, about 1.2 millimeters in diameter, are laid. After spawning, the coloration of the fish becomes somewhat lighter again. About 36 hours later one can already clearly

Courtship over the spawning site.

The spawning rock is cleaned.

The female attaches the eggs to the substrate.

The female fans the clutch.

The larvae are kept in depressions.

The parents with the school of fry.

Open spawners, spawning series of *Aequidens dorsiger*.

discern the development of the embryos in the eggs, which now have become visibly darker.

At temperatures of about 26° C, the larvae are sufficiently developed that they can leave the egg membrane after about 41 hours. Here too, the female usually sucks each individual larva from the egg membrane and places it in a prepared depression. The parents now change color once more. Both fish are now dark again. It now takes about another five days before the fry become free-swimming. They should be fed with the finest newly hatched *Artemia* or *Cyclops*.

Comments: In its natural habitat, the Red-breasted Cichlid is called "bobo" by the natives, which means "the dumb one." This refers to the minimal shyness of these fish, which one can easily catch by hand in the wild as well as in larger aquaria, once they have become well acclimated.

Genus *Anomalochromis* Greenwood, 1985

Explanation of the scientific name: Refers to the unusual nature of the gill cover-lower mandible lateral-line canal and to the former name for cichlids "chromides."

anomalos (Gk.) = not even, not the same; *chroma* (Gk.) = color.

Original description: Greenwood, P. H. (1985): The generic status and affinities of *Paratilapia thomasi* Blgr 1915 (Teleostei, Cichlidae). *Bull. British Mus. Nat. Hist. (Zool.),* vol. 49 (2), p. 257–272.

Distribution: Coastal rivers in Guinea, Sierra Leone, and Liberia (Africa).

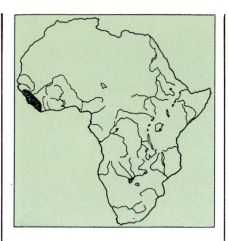

Number of species: So far only one species.

Generic characters: The dorsal fin has 14 spines and 9 to 10 soft rays; the anal fin has 3 spines and 7 to 8 soft rays. The fish have 24 to 26 scales in the lateral-line series. The scales on the upper lateral line are separated by a distance of one large and one small scale from the base of the dorsal fin.

Total length: 65 millimeters.

Type species: *Anomalochromis thomasi* (*Paratilapia thomasi* Boulenger, 1915).

Comments: Regan (1922) placed this species in the genus *Pelmatochromis*. In 1980 it was classified by Wilson and Loiselle in the genus *Hemichromis*.

Anomalochromis thomasi (Boulenger, 1915) African Butterfly Cichlid

Explanation of the scientific name: Refers to the discoverer of this species, N. W. Thomas.

Other names in the literature:
Paratilapia thomasi Boulenger, 1915
Pelmatochromis thomasi Regan, 1922
Hemichromis thomasi Wilson and Loiselle, 1980

Original description: Boulenger, G. A. (1915): Description of new freshwater fishes from Sierra Leone. *Ann. Mag. Nat. Hist.,* vol. 15 (8), p. 202–204.

Distribution: Guinea, Sierra Leone, and Liberia.

Habitat: Near the coast in small, flowing bodies of water with soft, clear water in rain forests. Prefers riparian sites with a rocky substrate and overhanging vegetation.

First importation into Germany: 1962 by Roloff (Karlsruhe).

Characters:
fins: D XIV/9–10, A III/7–8
scales: mLR 24–26
total length: male up to 65 millimeters; female up to 45 millimeters.

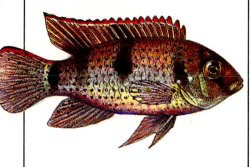

Sex differences: Relatively easy to differentiate only in older specimens. Females then have a more compact form than do males, and can also be recognized readily by the very light belly and breast areas.

Aquarium keeping: As in *Microgeophagus ramirezi*.

Spawning: Without problems. The fish spawn at almost any water hardness and at a water temperature of 24 to 28° C. Up to 700 fry are possible from one brood.

Anomalochromis thomasi guarding their eggs. The pair faced the camera to observe the movements going on in the vicinity of their spawn. **Opposite page:** *Cichlasoma nigrofasciatum* parent shepherding free-swimming fry.

Genus *Cichlasoma* Swainson, 1839

Explanation of the scientific name: Refers to the bright coloration of the fishes.

kichle (Gk.) = colorful bird, thrush; *soma* (Gk.) = body.

Original description: Swainson, W. (1839): The natural history of fishes, amphibians and reptiles or monocardian animals. II.

Distribution: Southern North America as far as the southern tributaries of the Rio Paraná in southern South America.

Number of species: So far over 100 species are known.

Generic characters: Typical cichlid form. In contrast to the genus *Aequidens*, the *Cichlasoma* species have more than three spines in the anal fin. *Aequidens* species possess three spines.

Total length: 100 to 700 millimeters.

Type species: *Cichlasoma punctatus* (*Labrus punctatus* Bloch).

Comments: Only a small portion of the *Cichlasoma* species have gained entry to the aquarium hobby. Many species are unsuitable for aquarium keeping because of their size. They are popular food fishes in their area of occurrence.

Cichlasoma nigrofasciatum
(Günther, 1869)
Convict Cichlid

Explanation of the scientific name: Refers to the black vertical stripes.

niger (L.) = black; *fasciatus* (L.) = striped, equipped with bands.

Other names in the literature: *Heros nigrofasciatus* Günther, 1869

Astronotus nigrofasciatum Eigenmann.

Original description: Günther, A. C. L. G. (1869): Account of the fishes of the states of Central America, based on collections made by Capt. J. M. Dow, F. Godman and V. Salvin. *Trans. Zool. Soc., London*, 6, p. 452 (*Heros nigrofasciatus*).

Distribution: Lake Amatitlan and Lake Atitlan in Guatemala.

First importation into Germany: 1934 by J. P. Arnold (Hamburg).

Characters:

fins: D XVII-XVIII/8–9, A VIII-X/6–8

scales: mLR 28–31

total length: male up to 100 millimeters; female up to 80 millimeters.

Sex differences: Males are darker colored than females and have thread-like elongated dorsal and anal fins.

Aquarium keeping: The fish do well in community tanks, where they will also spawn. Since the fish are very aggressive during brood care, however, keeping a pair in a species tank is more suitable. No particular demands with respect to water hardness; water temperature of 22 to 28° C.

Spawning: The fish spawn readily on a smooth and as level as possible substrate that is somewhat screened by fairly large plant leaves. The clutch can consist of up to 800 eggs. The brood is tended by both parents, whereby the female, however, devotes more attention to direct brood care, while the male takes charge of territorial defense. The larvae hatch after about 60 hours and are brought to previously prepared depressions. After another four days the fry become free-swimming and are led by both parents.

Comments: There is also a yellow mutation of this species.

Genus *Crenicara* Steindachner, 1875

Explanation of the scientific name: Refers to the serrated edges of the gill covers.

crenatus (L.) = provided with notches; *acara* = from the native name for cichlids.

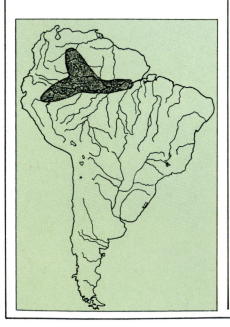

Original description: Steindachner, F. (1875): Beitrage zur Kenntnis der Chromiden des Amazonenstromes. *Sber. Akad. Wiss.,* Wien, 71, p. 99.

Distribution: River system of the upper Rio Negro, state of Amazonas (Brazil).

Number of species: So far four species are known. One species has not yet been described scientifically.

Generic characters: The edges of the gill covers are finely toothed. The dorsal fin has 14 to 17 spines as well as 6 to 9 soft rays; the anal fin has 3 spines and 5 to 8 soft rays. The fish exhibit, depending on mood, a checkerboard pattern in the body markings.

Total length: 70 to 120 millimeters.

Type species: *Crenicara punctulata* (*Acara punctulata* Günther, 1863: described as a new species by Steindachner, who was unaware of the prior description).

Crenicara filamentosa
Ladiges, 1958
Checkerboard Cichlid

Explanation of the scientific name: Refers to the pointed tail fin.

filamentosus (L.) = thread-like appendages.

Other names in the literature: *Dicrossus* sp. Fernandez-Yepez, 1969.

Original description: Ladiges, W. (1958): Bermerkungen zu einigen Neuimporten. *Die Aquar. und Terr.-Zschr.* (*DATZ*), 11, p. 203–204 (*Crenicara filamentosa*).

Distribution: The author collected them in small tributaries of the upper Rio Negro; apparently also found in tributaries of the upper Rio Orinoco.

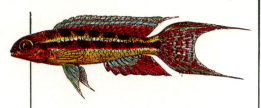

Habitat: Small, slow-flowing bodies of water with still-water zones in the rain forest; substrate slightly muddy and covered with small branches and leaves. Clear water with virtually no water hardness and *p*H values of between 4.8 and 6.

First importation into Germany: 1958 by Tropicarium Frankfurt (Main).

Characters:
fins: D XIV-XV/6–8, A III/6
scales: mLR 27–28
total length: male up to 90 millimeters; female up to 55 millimeters.

Sex differences: Males are more slender and colorful than females and have elongated tips of the tail fin. As a rule, females exhibit only a light-dark checkerboard pattern or a dark longitudinal band. Her tail fin is rounded off.

Aquarium keeping: As large as possible aquaria should be used. The tanks should be at least 80 centimeters long and should have a volume of at least 80 liters. It is possible to keep them over a fairly long period of time even in relatively hard water if the carbonate hardness does not exceed 7° dH. To be on the safe side, however, if possible the fish should be kept in water with no more than 10° dH of total hardness and a carbonate hardness of 2° to 3° dH. Too hard water could be the reason for many failures in the keeping of these splendid fish. In my experience, however, fairly frequent water changes appear to be considerably more important.

One plants the aquarium with fairly large *Nomaphila* and in the foreground with several bushy *Echinodoris horizontalis*. It is suitable to populate a species tank with one or two males and two to

Crenicara filamentosa male in the typical pre-spawning mode of pinched-in fins.

Crenicara filamentosa female guarding her spawn.

A closeup of the *Crenicara filamentosa* eggs.

four females as well as four to six smaller tetras that prefer the upper and middle water levels as secondary fishes. The False Rummy Nose Tetra (*Petitella georgiae*), for example, is suitable for this purpose. The fish then quickly lose their shyness, and for spawning attempts a certain hostile factor is then present. The water temperature can be between 24 and 30° C.

Spawning: The fish remain in the keeping tank. Once one has observed that the female's light ventral fins have turned a reddish color (only virginal females exhibit light ventral fins) or if one has noticed the clear presence of spawn in the female (very light belly area), one replaces the water in the aquarium with water with no more than 3° dH total hardness and no carbonate hardness. Through filtering over peat one should already have lowered and stabilized the *p*H beforehand. Then the *p*H cannot fall as rapidly into the dangerous range of less than 4.

If a female is ready to spawn, the male no longer exhibits the typical checkerboard pattern in its presence, but rather an uninterrupted dark longitudinal stripe. In a diagonal position (head down) the male then swims more and more often to the plants and here turns even a darker color. I was able to observe repeatedly a behavior that led me to assume that the male was almost ready to perish. In so doing it swam reeling and tottering with fins pressed to its body across the upper side of a horizontally positioned plant leaf. If the female approached, however, it suddenly spread all of its fins again and swam around her jerkily. During this time all other females and males were chased from the territory quite violently. It is therefore advisable either to remove the other fish from the tank or to provide good hiding places (resinous bog roots, coconut shells, rock structures).

In this phase the female also exhibits an uninterrupted dark longitudinal band in place of the checkerboard pattern. It now cleans a horizontally positioned leaf, although usually not as intensively as other dwarf cichlids clean the future spawning site. The short, somewhat dark spawning tube of the female is now visible. The female swims more and more often to the male, which is close by, and nudges it in the side. Later both mates stay only on the leaf and begin to spawn. This usually takes place in the late evening hours on the upper side of the leaf. Others of their species are driven energetically from the vicinity of the spawning site during this time; the secondary fishes, however, are left in peace, even if they venture quite near the clutch. Spawning itself lasts about a half hour. In the process up to 150 light amber-colored eggs are laid. After spawning, the female alone tends the clutch and chases the male from the immediate vicinity of the spawning site. Depending on water temperature, the larvae hatch 48 to 60 hours after the eggs are laid. The female gathers the larvae among plants or roots, but constantly moves them to fresh beds. The fry become free-swimming seven to eight days after the eggs are laid.

Under unfavorable conditions, females often eat the eggs only a few hours after spawning. After several failures one should therefore remove the eggs from the female and hatch them in a separate tank.

Crenicara maculata
(Steindachner, 1875)
Checkerboard Cichlid

Explanation of the scientific name: Refers to the checkerboard markings.
maculatus (L.) = spotted, speckled.
Other names in the literature:
Dicrossus maculatus
Steindachner, 1875
Crenicara praetoriusi Ahl, 1936

Crenicara maculata female. Imported into Germany in 1934. Photo by Harald Schultz (1960).

Original description: Steindachner, F. (1875): Beitrage zur Kenntnis der Chromiden des Amazonenstromes. *Sber. Akad. Wiss.,* Wien, 71, p. 102–106 (*Dicrossus maculatus*).

Distribution: Steindachner (1875) gives Lago Maximo at José Assu, the tributaries of the Amazon at Tonantins, the Rio Tajapuru, and the Rio Javari as collecting sites, whereas Ahl (1936) cites the Igarapé Irura-Mapiry. All data on the range speak for a relatively wide distribution along the Amazon; therefore, it is surprising that the species has not been imported again since 1950.

Habitat: Apparently similar to *Crenicara filamentosa.*

First importation into Germany: 1934, apparently only females; 1950 to Hamburg.

Characters:
fins: D XIV/9, A III/7
scales: mLR 26
total length: male up to 100 millimeters, female up to 60 millimeters.

This species is supposed to differ visually from *Crenicara filamentosa* principally through the single-tipped tail fin.

Coloration according to Ladiges (1951): Dorsal, anal, and tail fins with rows of wine-red spots and borders. Ten to twelve wine-red vertical bands on the tail fin. Bright blue and orange-red longitudinally striped ventral fins, which reach to the middle of the anal fin when pressed to the body.

This coloration also applies to *Crenicara filamentosa,* so that a certain scepticism is certainly appropriate. All of the pictures known to me show either females of *Crenicara filamentosa* (photographs) or males (drawings), which are confusingly alike except for the tail fin of *Crenicara filamentosa.*

Sex differences: Male with oval to lanceolate tail fin; dorsal and anal fins are tapered to points. Female has only a dark-light checkerboard pattern; the tail fin is shorter.

Aquarium keeping: As in *Crenicara filamentosa.*

Spawning: Praetorius (1935) writes that he obtained numerous offspring with one male and three females. The female cares for the fry alone.

According to Ladiges (1951), the fish spawn in a flowerpot in the usual manner of cichlids. The number of offspring was small. Based on his results, it must be a cavity spawner, in contrast to the other *Crenicara* species.

Comments: This species has not been imported again in the last forty years, even though it concerns a very attractive species whose localities are known.

This famous photograph taken by the late Gene Wolfsheimer in the 1950's was the first photo of a pair of *Crenicara maculata*. The male is the uppermost fish. This fish is widely dispersed but, according to Dr. Axelrod who collected them with Harald Schultz and alone, they are found in very shallow water completely choked with algae and growing plants, with a muddy bottom that precludes being walked on, thus making it very difficult to collect the fish alive in commercial quantities. Dr. Axelrod told the author that he found many of them outside the town of Humaita on the Rio Madeira, Brazil.

Crenicara punctulata
(Günther, 1863)
Spotted Checkerboard Cichlid

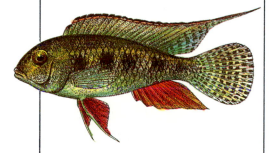

Explanation of the scientific name: Refers to the conspicuous dark spots along the midline of the body.
punctulatus (L.) = provided with small spots.

Other names in the literature:
Acara punctulata Günther, 1863
Crenicara elegans Steindachner, 1875
Aequidens madeirae Fowler, 1913
Aequidens hercules Eigenmann and Allen, 1942
Crenicara punctata Staeck, 1974.

Original description: Günther, A. C. L. G. (1863): Catalogue of the fishes in the British Museum. *Ann. Mag. Nat. Hist.,* 12, p. 441 (*Acara punctulata*).

Distribution: According to Kullander (1978), who summarized the most important localities, along the Amazon as far as its headwaters, as well as in Guyana and in a tributary of the Rio Madeira in the State of Rondonia (Brazil).

Habitat: Smaller rivers with moderate current and overgrown riparian zones; muddy substrate.

First importation into Germany: 1975.

Characters:
fins: D XVI/7, A III/5
total length: male up to 120 millimeters; female up to 80 millimeters.

Sex differences: The large fins of the male are all bluish colored; the head and breast area are yellow-gold. The anal, dorsal, and ventral fins are tapered to points. The female's fins are smaller in area and are rounded off. Her ventral and anal fins are reddish colored.

Thanks to aquarium observations by fanciers, a sex change not previously encountered in any other cichlid species was confirmed, which was described by Ohm (1980). Juveniles all exhibit a distinct female-like coloration. In a small group of fish of this species, in each case only one male develops. If one removes the male from the tank, another female develops into a male and so on. All former females become males in the final stage.

Aquarium keeping: Fairly large, well-planted aquaria are necessary for these fish. They can be kept in almost any tap water. Frequent water changes are the only important considerations. It is suitable to keep these fish with tetras in order to decrease their shyness. It is also possible to keep them in a community tank. It is best to buy three or four juveniles so that one can observe the transformation of one of the females into a male over the course of time.

Spawning: Before attempting to provide suitable spawning conditions for the fish, it is essential that they be sexually mature and that one of them has changed completely into a male. It is not difficult to recognize ready-to-spawn females by the yellow-gray coloration, on which the checkerboard pattern is still barely visible. In addition to a rich yellow-gold coloration of the head and belly areas, courting males exhibit a dark longitudinal stripe that forms from the middle row of spots.

Once we recognize these characters, it is suitable to siphon out two thirds of the water and to replace it with completely desalinated water that has been filtered through peat. Further preparations are not necessary. After courtship, the female searches for a horizontally positioned leaf, which if possible should be shaded by other leaves. This is then intensively cleaned by the female. During this time the

Crenicara punctulata ♂.

Crenicara punctulata ♀.

Cichlasoma nigrofasciatum.

male always stays close by. After the female has laid the first 30 to 50 eggs on the leaf, the male swims up and fertilizes them. I observed that if there was a disturbance in front of the tank, the male swam only very briefly and only a few times over the clutch. Therefore, as far as possible one should leave the fish alone during spawning.

The male is driven off by the female immediately after spawning. The female's brood care is extremely intense. After about 60 hours the female sucks the larvae from the egg membranes and places them in a depression in the substrate. The fry become free-swimming after an additional five days.

Comments: The sex change from female to male also takes place in females that have already spawned repeatedly. According to Ohm (1980), the sex change takes place in conformance with a social hierarchy. If no male is present, the dominant female thus always develops into a new male.

**Genus *Etroplus*
Cuvier and Valenciennes, 1830**

Explanation of the scientific name: Refers to the stiff anal fin rays.

etron (Gk.) = belly; *hoplon* (Gk.) = weapon.

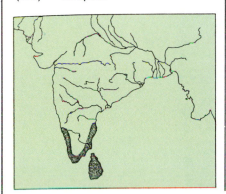

Original description: Cuvier, G., and Valenciennes, A. (1830): Histoire naturelle des poissons. 5, p. 486, Levrault, Paris.

Distribution: Southern India and Sri Lanka.

Number of species: So far three species are known.

Generic characters: The fishes possess a large number (about 12 to 15) of spines in the anal fin.

Total length: 90 to 400 millimeters.

Type species: *Etroplus suratensis* (*Chaetodon suratensis* Bloch, 1790), which was, nevertheless, named *Etroplus meleagris* by the first describers, Cuvier and Valenciennes (1830).

Comments: So far only one species has become established in aquaria. The other two species are either too large (*Etroplus suratensis*) or have not yet been imported (*Etroplus canarensis*).

Etroplus maculatus (Bloch, 1795)
Orange Cichlid

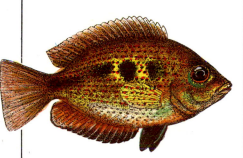

Explanation of the scientific name: Refers to the black spots on the sides of the body.

maculatus (L.) = speckled, spotted.

Other names in the literature:
Chaetodon maculatus Bloch, 1795
Glyphisodon kakaitsel Lacépède, 1802
Etroplus coruchi Cuvier and Valenciennes, 1830.

Original description: Bloch, M. E. (1785–1795): Naturgeschichte der auslandischen Fische. Berlin, vol. 9, p. 244 (*Chaetodon maculatus*).

Distribution: Southern India and Sri Lanka.

First importation into Germany: 1905 by Reichelt (Berlin).

Characters:
fins: D XVII–XX/8–10, A XII–XV/8–9, P 15–16
scales: mLR 35–37
total length: male up to 90 millimeters; female up to 80 millimeters.

Sex differences: Males have somewhat more intense coloration.

Aquarium keeping: As far as possible, one should use tanks with a length of at least 80 centimeters that are well planted in the background and decorated with resinous bog wood as well as rocks. No particular demands with respect to water hardness; water temperature between 22° and 28° C. The fish are best kept in pairs. The often mentioned instruction that it is necessary to add some salt to the water is not correct in my experience. Much more important are frequent water changes.

Spawning: The fish prefer to spawn on somewhat concealed sites, on or against a rock. After spawning, the clutch is principally cared for by the female. The hatched larvae are kept in depressions. After the fry become free-swimming they are led by both parents. The fry are fed with newly hatched *Artemia* or *Cyclops*. They are also said to feed on a slimy secretion that is produced on the bodies of the parents.

Comments: For several years there has also been a yellow mutation of this fish on the market.

Genus *Hemichromis* Peters, 1857

Explanation of the scientific name: Refers to the former name for cichlids, "chromides."
hemi (Gk.) = half; *chroma* (Gk.) = color.

Original description: Peters, W. C. H. (1857): Neue Chromidengattung (*Hemichromis*). *Monatsber. Akad. Wiss. Berlin,* p. 403.

Distribution: Western and northwestern Africa.

Number of species: According to the revision by Loiselle (1979), eleven species are known.

Generic characters: These are beautifully colored, small to medium-sized cichlids with an elongated, laterally compressed form. The dorsal fin has 13 to 15 spines and 9 to 13 soft rays, and the anal fin has 3 spines and 7 to

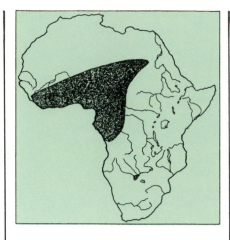

13 soft rays. The fins, which are tapered to points, are only slightly elongated.
Total length: 70 to 300 (?) millimeters.
Type species: *Hemichromis fasciatus* Peters, 1857.

Hemichromis bimaculatus Gill, 1862
Jewel Fish

Explanation of the scientific name: Refers to the dark spots exhibited on the body of the fish.
bimaculatus (L.) = provided with two spots.
Other names in the literature:
Hemichromis guttatus Günther, 1862
Hemichromis letourneuxii Sauvage, 1880
Hemichromis saharae Sauvage, 1880
Hemichromis rolandi Sauvage, 1881
Hemichromis fugax Payne and Trewavas, 1976.
Original description: Gill, T. (1862): On the West African genus *Hemichromis* and descriptions of new species in the

Etroplus maculatus with free-swimming fry.

Museum of the Academy and the Smithsonian Institution. *Proc. Acad. Nat. Sci. Philadelphia*, p. 137 (*Hemichromis bimaculatus*).

Distribution: Sierra Leone, Guinea, and Liberia.

Habitat: Small water courses in the coastal region with clear, soft, and slightly acidic water.

First importation into Germany: 1907 by the Vereinigten Zierfischzuchtereien Conradshohe.

Characters:

fins: D XIV-XV/10–12, A III/8–9, P 14–15

scales: mLR 26–28

total length: male up to 120 millimeters; female up to 100 millimeters.

Hemichromis lifalili with her free-swimming fry.

Sex differences: Males are mainly light red in the throat area; the dorsal and anal fins are more pointed. Females are light red except for the forehead area; the fins are not as large in area.

Aquarium keeping: Fairly large aquaria are necessary. The fish make no particular demands with respect to water hardness.

Spawning: Possible without special preparations. The fish spawn on a flat rock. The brood is tended by both parents.

Comments: Up to now this species has only rarely been imported. Most of the fish that have previously gone by this name were actually *Hemichromis guttatus*.

Hemichromis cerasogaster
(Boulenger, 1899)

Explanation of the scientific name: Refers to the cherry-red belly of the fish.

cerrasum (L.) = cherry; *gaster* (Gr.) = belly.

Other names in the literature: *Paratilapia cerasogaster* Boulenger, 1899

Pelmatochromis cerasogaster Regan, 1922.

Original description: Boulenger, G. A. (1899): Materiaux pour la faune du Congo. Poissons nouveaux du Congo. V. Partie: Cyprins, Silures, Cyprinodonts, Acanthopterygiens. *Ann. Mus. Congo, Zool.* ser. 1, p. 118 (*Paratilapia cerasogaster*).

Distribution: Endemic to Lake Mai-Ndombe (Lake Leopold II) in Zaire.

Characters:

fins: D XII-XV/10–12, A III/7–8

scales: mLR 25–27

total length: male up to 90 millimeters; female up to 80 millimeters.

Sex differences: Males beige, females in breeding coloration cherry red.

Aquarium keeping: As in *Hemichromis bimaculatus*.

Spawning: As *Hemichromis bimaculatus*.

Hemichromis cristatus
Loiselle, 1979
Red Eye-Spot Cichlid

Explanation of the scientific name: Refers to the presence of small dental protuberances on the lower pharyngeal teeth.

cristatus (L.) = having a crest.

Original description: Loiselle, P. V. (1979): A revision of the genus *Hemichromis* Peters, 1858 (Teleostei: Cichlidae). *Ann. Mus. Royal de l'Afrique centrale Sci. Zool.*, ser. 8, nr. 228, p. 54–59 (*Hemichromis cristatus*).

Distribution: Southeastern Guinea (Grandes Chutes River, Banama River, Frig ia be River, Kenini River), Sierra Leone, western Ghana, Nigeria (Ogba River), and southern Liberia.

Habitat: Small rivers, principally in forested regions; clear water with almost no hardness at a *p*H of about 5.

First importation into Germany: 1985 by Bleher (Aquarium Rio).

Characters:

fins: D XIV-XV/11, A III/8–9

scales: mLR 26–27

total length: male up to 90 millimeters; female up to 80 millimeters.

Sex differences: Readily visible differences in coloration are present only in specimens in breeding condition. Males have a rich red-brown colored ventral half of the body. Females are an intense orange-red color.

Aquarium keeping: As in *Hemichromis bimaculatus*.

Spawning: As in *Hemichromis bimaculatus*.

Hemichromis cristatus ♂.

Hemichromis cristatus ♀.

Hemichromis bimaculatus.

Hemichromis lifalili on the clutch.

Hemichromis lifalili ♀, normal coloration.

Spawning series of *Hemichromis lifalili*.

This is probably the most colorful of all aquarium fishes. Unfortunately it has a nasty disposition when it thinks about spawning! This pair of *Hemichromis lifalili* are swimming around with their very large spawn.

Hemichromis guttatus
Günther, 1862

Explanation of the scientific name: Refers to the in partly teardrop-shaped black spot on the middle of the body and the black spot on the gill cover.
guttatus (L.) = speckled, as if by drops.
Other name in the literature: *Hemichromis bimaculatus* Boulenger 1915.
Original description: Günther, A. C. L. G. (1862): Catalogue of the fishes in the British Museum. vol. IV. *British Mus. (Natural History)*, London, p. 275 (*Hemichromis guttatus*).
Distribution: Western to central Ghana, as well as in coastal regions from Togo to Cameroon.
Habitat: The most diverse types of bodies of water, chiefly in sites with heavy aquatic plant growth.
Characters:
fins: D XIV–XV/9–11, A III/8–9
scales: mLR 25–28
total length: male up to 120 millimeters; female up to 100 millimeters.
Sex differences: Males in breeding coloration violet-red, females cherry-red.

Aquarium keeping: As in *Hemichromis bimaculatus*.
Spawning: As in *Hemichromis bimaculatus*.
Comments: This species went by the name *Hemichromis bimaculatus* for a long time in the aquarium hobby.

Hemichromis lifalili
Loiselle, 1979

Explanation of the scientific name: Refers to the local name of these fish in the region of Lake Tumba in Zaire.
Other names in the literature: *Hemichromis bimaculatus II*?
Original description: Loiselle, P. V. (1979): A revision of the genus *Hemichromis* Peters, 1858 (Teleostei: Cichlidae). *Ann. Mus. Royal de lAfrique centrale, Scie. Zool.*, ser. 8, nr. 228, p. 113–121 (*Hemichromis lifalili*).
Distribution: Congo (Zaire) River system in the region around Kasai and Shaba (Ruki River, Lake Tumba, Lake Yandja) as well as in the upper course of the Ubangi River (Central African Republic).
Habitat: The fish occur in virtually all types of bodies of water. As a rule, the water in the collecting regions is very soft (total hardness up to 4° dH) and slightly acidic (*pH* between 5 and 6.2). The fish prefer to stay in riparian regions with a rocky substrate or dense riparian growth.
First importation into Germany: 1968.
Characters:
fins: D XIII–XV/10–12, A III/8–9
scales: mLR 25–27
total length: male up to 100 millimeters; female up to 80 millimeters.
Sex differences: In breeding coloration, there are well-defined differences in color: males have a dark coloration tending to lilac; females are bright cherry red.
Aquarium keeping: As in *Hemichromis bimaculatus*.
Spawning: As in *Hemichromis bimaculatus*.

Hemichromis paynei
Loiselle, 1979
Payne's Red Cichlid

Explanation of the scientific name: Refers to the ichthyologist Dr. A. Payne.
Original description: Loiselle, P. V. (1979): A revision of the genus *Hemichromis* Peters, 1858 (Teleostei: Cichlidae). *Ann. Mus. Royal de l'Afrique centrale, Sci. Zool.*, ser. 8, nr. 228, p. 59–65 (*Hemichromis paynei*).
Distribution: Chiefly in the region of estuaries in Guinea, Sierra Leone (Lake Kwarko), and Liberia (St. John River at Hartford, St. Poul River Lagoon).
First importation into Germany: 1981.
Characters:
fins: D XIII–XV/10–11, A III/7–8
scales: mLR 26–27
total length: male up to 110 millimeters; female up to 90 millimeters.

Sex differences: The sexes can be distinguished only by size, and during spawning time and when tending the brood. The male has a darker coloration, tending to lilac. The female is light red.
Aquarium keeping: As in *Hemichromis bimaculatus*.
Spawning: As in *Hemichromis bimaculatus*.

Hemichromis paynei female laying eggs.

Hemichromis paynei male.

Hemichromis paynei female with her free-swimming youngsters. If it is possible to raise fish alone with their parents, the sight of a large family is one of the most rewarding aspects of keeping dwarf cichlids . . . or any kind of fish. Most fishes, however, do not exhibit care of their offspring. Many fishes eat their own eggs and fry!

Genus *Nannacara*
Regan, 1905

Explanation of the scientific name: Refers to the size of the fish of the type species.
nanus (L.) = small, dwarf-like; *acara* = local name for cichlids in South America.

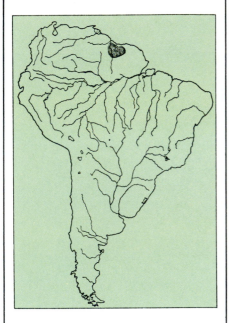

Original description: Regan, C. T. (1905): A revision of the fishes of the South American Cichlid genera *Acara*, *Nannacara*, *Acaropsis*, and *Astronotus. Ann. Mag. Nat. Hist.*, ser. 7, vol. 15, p. 344.
Distribution: Guyana, French Guiana.
Number of species: So far two species have been described.
Generic characters: Divided lateral line; the upper lateral line is located a half a scale row from the dorsal fin. The dorsal fin has 15 to 17 spines and 7 to 8 soft rays; the anal fin has 3 spines and 7 to 8 soft rays.
Total length: 80 to 100 millimeters.
Type species: *Nannacara anomala* Regan, 1905.

Nannacara anomala Regan, 1905
Golden-eyed Dwarf Cichlid

Explanation of the scientific name: Refers to the deviant placement of the lateral line.
anomalos (Gk.) = exceptional, deviant.
Other names in the literature:
Acara punctulata Günther, 1863
Nannacara taenia Regan, 1912.
Original description: Regan, C. T. (1905): A revision of the fishes of the South American Cichlid genera *Acara*, *Nannacara*, *Acaropsis*, and *Astronotus. Ann. Mag. Nat. Hist.*, ser. 7, vol. 15, p. 344–345 (*Nannacara anomala*).
Distribution: Guyana.
Habitat: In the most diverse types of bodies of water; prefers sites in which the plants and branches offer good concealment. As a rule, water is soft (total hardness of up to 8° dH) and slightly acidic (*p*H value of 5 to 6.4).
First importation into Germany: 1911 by Siggelkow (Hamburg).
Characters:
fins: D XVI/8, A III/8
scales: mLR 23–24
total length: male up to 80 millimeters; female up to 50 millimeters.
Sex differences: The male is intensely colored. The normal coloration of the female, in comparison, is plain. Only during the time of brood care is the conspicuous checkerboard markings perceived.

Aquarium keeping: One should keep this species in pairs in tanks at least 70 centimeters long. There absolutely must be numerous hiding places present. It is appropriate to keep them with

smaller tetras. No demands with respect to water hardness. The water temperature can lie between 22° and 28° C.
Spawning: For spawning, the same tank can be used as for keeping. It is recommended, however, that you place a flat rock in a sheltered site (under a few large-leaved aquatic plants), which will then be accepted as a spawning site. One can easily recognize ripe females by the light, swollen belly area. They are intensively courted by the males. Ready-to-spawn females then search for a suitable spawning site together with the male. The place on which the eggs will later be deposited is intensively cleaned by both fish. The female then glides over the cleaned site with her spawning tube and attaches the eggs. After the first row of eggs has been attached, the female swims to the side and the male glides over the eggs. In the process sperm is released and the eggs are fertilized. This is repeated until all of the eggs, a total of up to 300, have been released. During spawning, the female exhibits two dark longitudinal stripes. Spawning lasts about a half hour, after which the mates separate. In a fairly large tank, the male guards the territory at some distance from the spawning site, whereas the female takes charge of the direct brood care at the clutch. In this species, in particular, serious fighting can take place between the fish if the tank is too small and hiding places are lacking. The male can literally be hounded to death. It is therefore recommended in such a case that the male be removed from the tank immediately after spawning.
The larvae are peeled from the egg membranes by the female after 45 to 50 hours (depending on water temperature) and are placed in shallow depressions. Again and again the larvae are taken into the female's mouth and are cleaned by means of chewing movements. During the following five days the larvae do not always stay in the

Two male _Nannacara anomala_ fighting it out with a typical "jaw-lock," a common grip which results in no damage to either fish.

same depression, but, on the contrary, are transferred constantly. In the last phase of development to fry, the young already follow the appropriate body movements of the mother fish, which has now assumed the typical checkerboard markings. In so doing the young always hop short distances. Only when the yolk sac has been absorbed completely do they swim in a dense school around the female. In a large, well-furnished tank one can also observe that the male leads the fry now and then for a short time. It then exhibits virtually the same checkerboard pattern as the female.

Another favorite aggressive stance of _Nannacara anomala_ is the "fan-off." The two males display in front of each other, the same way they might display in front of a courting female.

Nannacara aureocephalus
Allgayer, 1983
Golden-Head Cichlid

Explanation of the scientific name: Refers to the gold color of the head region.
aureus (L.) = golden; _kephale_ (Gk.) = head.

Original description: Allgayer, R. (1983): _Nannacara aureocephalus_, espèce nouvelle de Guyane française (Pisces, Cichlidae). _Rev. Franc. Cichl._ 11, nr. 33, p. 13–16 and 21–24 (_Nannacara aureocephalus_).

Distribution: French Guiana, tributaries of the Mana near the settlement of Saut Sabbat on the National Road Nr. 1 and the tributary of the Comte near the settlement of Cacao.

Habitat: Streams and residual water holes in the rain forest

region. Slightly brownish water with a total hardness of up to 1° dH and a _p_H of about 6.

First importation into Germany: 1984.

Characters:
fins: D XV–XVII/7–8, A III/7–8
scales: mLR 23–24
total length: male up to 100 millimeters; female up to 60 millimeters.

Sex differences: Males are more conspicuously colored than the gray-brown females, in which a pattern of dark spots is suggested.

Aquarium keeping: As in _Nannacara anomala_.

Spawning: As in _Nannacara anomala_. However, fairly soft water with a total hardness of up to 10° dH should be used, at least with wild-caught specimens.

Genus *Microgeophagus*
Axelrod, 1971
Explanation of the scientific name: Refers to the smallness of the species and their relationship to another cichlid genus, *Geophagus*.

mikros (Gk.) = small;
Geophagus = genus of cichlids.

Original description: Axelrod, H. R. (1971): Breeding Aquarium Fishes, Book 2. T.F.H. Publ., Inc., New Jersey, U.S.A. p. 344–352.
map

Distribution: Bolivia, Venezuela, Colombia.

Number of species: So far two species are known.

Characters of the genus: The dorsal fin begins behind the gill cover edge (in the *Apistogramma* species in front of the gill cover edge). A small lobe is located on the upper part of the first gill arch.

Total length: 50 to 75 millimeters.

Type species: *Microgeophagus ramirezi* (*Apistogramma ramirezi* Myers and Harry, 1948).

Comments: The fishes of this genus went by the generic name *Apistogramma* for a long time, which even today is still sometimes used for them.

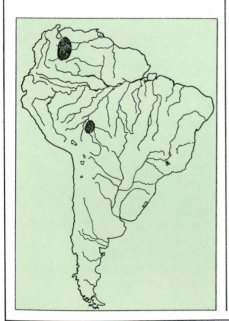

Microgeophagus altispinosa
(Haseman, 1911)
Bolivian Butterfly Cichlid

Explanation of the scientific name: Refers to the greatly elongated first ray of the dorsal fin.
altus (L.) = high; *spinosus* = spiny.

Other name in the literature:
Crenicara altispinosa Haseman, 1911.

Original description: Haseman, J. D. (1911): An annotated catalog of the Cichlid fishes collected by the expedition of the Carnegie Museum to Central South America, 1907–1910. *Ann. Carnegie Mus.,* Issue 7, p. 344–345 (*Crenicara altispinosa*).

Distribution: Rio Mamoré above and below the junction of the Rio Guapore, among other places at the town of Trinidad; river basin of the Rio Guapore at Santa Cruz (Bolivia); Rio Quizer near San Ramon, depression below Todos Santos (Bolivia); mouth of the Igarapé near Guajara-Mirim (Brazil).

First importation into Germany: 1984.

Characters:
fins: D XIII-XIV/9, A III/8, P 11–12
scales: mLR 26–29
total length: male up to 95 millimeters; female up to 80 millimeters.

Sex differences: Only present to a small degree in older specimens. Females produce a more compact impression.

Aquarium keeping: If possible, tanks should not be less than 80 centimeters long; dense planting except for open swimming space in the foreground.

Spawning: Possible in the keeping tank; nevertheless, the water should be soft (total hardness of up to 10° dH).

Comments: This species differs clearly in body form and size from *Microgeophagus ramirezi*. The fish frequently offered as *M. altispinosa* were in most cases unusually large *M. ramirezi* (strain imported from Singapore).

Microgeophagus ramirezi
(Myers and Harry, 1948)
Ramirez's Dwarf Cichlid

Explanation of the scientific name: Refers to M. P. Ramirez, who was the first to find this species.

Other names in the literature:
Apistogramma ramirezi Myers and Harry, 1948
Microgeophagus ramirezi Frey, 1957 (nomen nudem)
Pseudogeophagus ramirezi Hoedeman, 1969
Pseudoapistogramma ramirezi Axelrod, 1971
Papiliochromis ramirezi Kullander, 1977

Original description: Myers, G. S., and Harry, R. R. (1948): *Apistogramma ramirezi*, a Cichlid fish from Venezuela. *Proc. Calif. Zool. Club,* 1 (1), p. 1–8 (*Apistogramma ramirezi*).

Microgeophagus ramirezi spawning. ▲

The male has a longer second dorsal spine. ▲

The pair constantly circle each other during spawning, with the female depositing eggs and the male fertilizing them. ▲

Sometimes they follow each other so closely that their tails are in each other's face. ▲

The male often guards the eggs, during which time he develops a red belly. ▼

The male with his free-swimming fry. ▼

Distribution: Bodies of water in savannas northwest of the Rio Orinoco in Venezuela and Colombia; river system of the Rio Meta and Rio Vichade (Colombia).

Habitat: Clear, slow-flowing streams with muddy to sandy substrates and, in some cases, abundant vegetation. Water almost without hardness at pH values of about 5.5; water temperature about 30° C.

First importation into Germany: 1948 by Aquarium Hamburg.

Characters:

fins: D XIV-XV/9, A III/8, P 11–12

scales: mLR 26–29

total length: male up to 70 millimeters; female up to 50 millimeters.

Aquarium keeping: Considered as delicate, yet successful keeping apparently is dependent only on how high the concentration of harmful substances is in the water. Therefore, the water should be changed regularly. Fresh water works better than the always called for soft water.

Spawning: Presents no difficulties. The fish prefer to spawn on flat reddish rocks that are located somewhat in the shade of aquatic plants. Softer water brings better success in spawning. The greatest number of offspring known to me was about 500 fry from one clutch.

There are many color varieties of *Microgeophagus ramirezi*, from the dark, wild color shown above to pale yellow. The wild color form is by far the most attractive. The male, with his free-swimming fry, is in a very colorful phase.

Genus *Apistogramma* Regan, 1913

Explanation of the scientific name: Refers to the irregular lateral line.

apistos (Gk.) = irregular, nonuniform; *gramma* (Gk.) = mark, writing.

Original description: Regan, C. T. (1913): Fishes from the river Ucayali, Peru, collected by Mr. Mounsey. *Ann. Mag. Nat. Hist.,* ser. 8, vol. 12, p. 282.

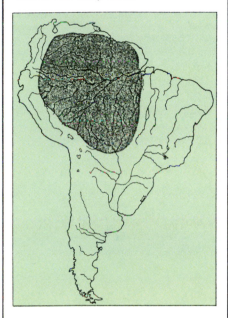

Distribution: Entire tropical South America.

Number of species: So far about 60 species are known.

Generic characters: Divided, displaced lateral line; three dark bands, more or less conspicuous depending on mood, that extend from the eye to the mouth, to the lower edge of the gill cover, and to the tail fin.

Total length: 30 to 120 millimeters.

Type species: *Apistogramma taeniata* (*Mesops taeniatus* Günther, 1862).

Comments: Up until 1906, various *Apistogramma* species were classified in the genera *Geophagus, Mesops,* and *Biotodoma.* Regan (1906) then combined them in the genus *Heterogramma.* Since the generic name *Heterogramma* was already used for a genus of beetles, Regan (1913) altered the generic name through the substitution of the first two syllables. *Hetero-* was replaced by *Apisto-*.

Apistogramma agassizii (Steindachner, 1875) Agassiz's Dwarf Cichlid

Explanation of the scientific name: Refers to Prof. H. R. L. Agassiz, who discovered the species and delivered it for description.

Other names in the literature:
Geophagus (Mesops) aggassizii Steindachner, 1875
Biotodoma agassizii Pellegrin, 1904
Heterogramma agassizi Haseman, 1911
Apistogramma agassizi Vandewalle, 1973

Original description: Steindachner, F. (1875): Beitrage zur Kenntnis der Chromiden des Amazonenstromes. Sber. Akad. Wiss. Wien, 71, p. 111–115 [*Geophagus (Mesops) agassizii*].

Distribution: The immense range of this species extends from the Rio Ucayali (Peru) over virtually the entire main river basin of the Amazon as far as Santarem (Brazil).

Habitat: The fish are chiefly found in rivers with rocky or sandy substrates close to the banks, where suitable shelters in the form of cavities (rocks, roots, or branches) are present. As a rule, the clear water is very soft and slightly acidic.

First importation into Germany: 1909 by Siggelkow (Hamburg).

Characters:
fins: D XV/7, A III/6, P 14
scales: mLR 23, Ltr 11–12
total length: male up to 100 millimeters; female up to 60 millimeters.

Sex differences: Males are conspicuously colored and have larger, elongated fins. Females have a yellowish ground color and a dark, almost black longitudinal stripe on the middle of the body that extends from the rear edge of the eye right into the tail fin.

Aquarium keeping: In order to display the full beauty of these fish, one should not keep them in a community tank, but in a species tank. It is nevertheless appropriate to keep two males and three to four females in a fairly large aquarium with a few small tetras. Since several males will set up their territories and the females also need smaller territories, dense planting as well as several cavities absolutely must be present. The aquarium should be at least 80 centimeters long and hold about 100 liters. This species can be kept readily even in fairly hard water; the only important thing is fairly frequent water changes.

Spawning: Soft water is needed to spawn these fish. The total hardness can be between 2° and 10° dH.

Comments: Because of the extensive range of this fish, this species exhibits a great range in variation in coloration. At present one distinguishes three basic forms: the blue (Manaus), yellow (Santarem), and red (Colombia, Peru) color forms. In addition, breeders have bred a large number of color varieties, which in part eclipse the natural color forms.

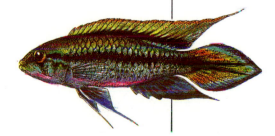

Apistogramma agassizi female beginning to lay her eggs. Only a few randomly laid eggs are visible.

A lovely male *Apistogramma agassizi*.

Apistogramma bitaeniata
Pellegrin, 1936
Double-Band Dwarf Cichlid

Explanation of the scientific name: Refers to the two dark longitudinal bands this species possesses. The second band, under the always visible longitudinal band, is always only suggested and is dependent on mood.
bi- (L.) = twice, double; *taenia* (L.) = band.

Other names in the literature:
Apistogramma pertense var. *bitaeniata* Pellegrin, 1936
Apistogramma sweglesi Meinken, 1961
Apistogramma klausewitzi Meinken, 1964
Apistogramma kleei Meinken, 1964

Original description: Pellegrin, J. (1936): Un poisson d'aquarium nouveau du genre *Apistogramma*. Bull. Soc. Natn. Acclimat. Prot. Nat., Paris, 83, p. 56–58 (*Apistogramma pertense* var. *bitaeniata*).
Distribution: Upper Amazon River basin (Peru and Brazil), Departemento Amazonas (Colombia), vicinity of Leticia (border area between Peru and Brazil).
Habitat: Smaller water courses with a sandy substrate near the banks; extremely mineral-poor, soft, and slightly acidic water.
First importation into Germany: 1960 by Dr. E. Schmidt-Focke (Bad Homburg).
Characters:
fins: D XV-XVI/6–7, A III/5–6, P 14, V I/5
scales: mLR 23–24

A male *Apistogramma bitaeniata* in breeding colors.

A pair of *Apistogramma bitaeniata* guarding their eggs.

total length: male up to 80 millimeters; female up to 60 millimeters.

Sex differences: Males are more brightly colored than females, with elongated fins and a forked tail fin; females with a yellow ground color and rounded-off tail fin.

Aquarium keeping: As in *Apistogramma agassizii.*

Spawning: As in *Apistogramma agassizii,* except softer water lower in minerals should be used.

Comments: A yellow and a red color variety of this species are also known.

Apistogramma borelli
(Regan, 1906)
Yellow Dwarf Cichlid

Explanation of the scientific name: Refers to Dr. A. Borelli, the discoverer of this species.

Other names in the literature:
Heterogramma borellii Regan, 1906
Heterogramma ritense Haseman, 1911
Heterogramma rondoni Miranda Ribeiro, 1918
Apistogramma aequipinnis Ahl, 1939
Apistogramma reitzigi Ahl, 1939

Original description: Regan, C. T. (1906): A revision of the South-American Cichlid genera *Retroculus, Geophagus, Heterogramma,* and *Biotoecus. Ann. Mag. Nat. Hist.,* ser. 7, vol. 17, p. 63–64 (*Heterogramma borellii*).

Distribution: Southern Brazil (Mato Grosso region, La Plata region), Argentine river basin of the Rio Paraguai (Rio Santa Rita), and junction of the Rio Paraguai and the Rio Paraná.

First importation into Germany: 1936 by H. Rose (Hamburg).

Characters:
fins: D XVI/5–6, A III/6–7
scales: mLR 21
total length: male up to 70 millimeters; female up to 45 millimeters.

Sex differences: Males have elongated fins and are more conspicuously colored than the yellow-colored females.

Aquarium keeping: As in *Apistogramma agassizii.*

Spawning: As in *Apistogramma agassizii.*

Comments: This species was called *Apistogramma reitzigi* for a long time in the aquarium hobby after it was described subsequently by Ahl (1939). This name is occasionally still used even today.

Various color varieties of *Apistogramma borellii* are known.

Apistogramma brevis
Kullander, 1980

Explanation of the scientific name: Refers to the small size of this species.

brevis (L.) = short, small.

Original description: Kullander, S. O. (1980): A taxonomical study of the genus *Apistogramma* Regan, with a revision of Brazilian and Peruvian species (Teleostei: Percoidei: Cichlidae)). *Bonner Zool. Monograph.,* nr. 14, p. 107–111 (*Apistogramma brevis*).

Distribution: Brazil (Lago Penera, Rio Uaupés).

First importation into Germany: So far not imported alive.

Characters:
fins: D XV-XVI/6, A III/5–7
scales: mLR 22–23
total length: male 55.4 millimeters; female 36 millimeters (type specimen, after Kullander, 1980).

Apistogramma cacatuoides
Hoedeman, 1951
Cockatoo Dwarf Cichlid

Explanation of the scientific name: Refers to the elongated dorsal fin rays that are raised in excitement as in the cockatoo.

cacatua (Sp.) = cockatoo; *eidos* (Gk.) = appearance; *-oides* (L.) = similar.

Other names in the literature:
Apistogramma marmoratus Dunker, 1960
Apistogramma borelli Meinken, 1961
Apistogramma U2 Innes, 1966.

Original description: Hoedeman, J. J. (1951): Notes on the fishes of the Cichlid family I. *Apistogramma cacatuoides* sp. n. *Beaufortia, Zool. Mus. Amsterdam.* nr. 4, p. 1–4 (*Apistogramma cacatuoides*).

Distribution: Departemento Amazonas (Colombia); Rio Yavari, Lago Matamata, vicinity of Pucallpa (Peru).

Habitat: Smaller, shallow rivers and accumulations of water in the rain forest; bottom covered with a layer of leaves and branches. Medium-hard water with a pH of between 7 and 8; water temperature 25° C.

First importation into Germany: 1950 by Hoedeman (Netherlands).

Characters:
fins: D XV-XVII/5–8, A III-IV/5–7
scales: mLR 22–24
total length: male up to 80 millimeters; female up to 50 millimeters.

Sex differences: Males more strikingly marked than females,

A male *Apistogramma cacatuoides*.

with elongated forward dorsal fin rays and fins tapering to points. Females yellow-gray with dark longitudinal stripes.

Aquarium keeping: As in *Apistogramma agassizii*; a very robust species that does well even under relatively unfavorable aquarium conditions. Males are very aggressive toward one another.

Spawning: As in *Apistogramma agassizii*.

Comments: This species went by the name of *Apistogramma borellii* in the aquarium hobby after the description by Meinken (1961), a name that is still often used today.

Today several strains (color varieties) can be found in aquaria.

Apistogramma caetei
Kullander, 1980
Rio Caete Dwarf Cichlid

Explanation of the scientific name: Refers to the main river of the locality of the type specimen, the Rio Caeté in Brazil.

Other name in the literature:
Heterogramma taeniata Haseman, 1911.

Original description: Kullander, S. O. (1980): A taxonomical study of the genus *Apistogramma* Regan, with a revision of Brasilian and Peruvian species (Teleostei: Percoidei: Cichlidae:). *Bonner Zool. Monograph.*, nr. 14, p. 76–79 (*Apistogramma caetei*).

Distribution: Igarapé at Braganca, Rio Apeu in the state of Para (Brazil).

First importation into Germany: 1980.

Characters:
fins: D XV/6–7, A III/6–7
scales: mLR 23
total length: male up to 60 millimeters; female up to 40 millimeters.

Sex differences: Males are more colorful than the yellow-gray to yellow females and their dorsal fin is tapered to a point.

Aquarium keeping: As in *Apistogramma agassizii*.

Spawning: As in *Apistogramma agassizii*.

Apistogramma commmbrae
(Regan, 1906)
Corumba Dwarf Cichlid

Explanation of the scientific name: Refers to the locality Corumba in Brazil, which, however, Regan apparently misread in Eigenmann's records, which led to this descrepancy.

Other names in the literature:
Heterogramma commmbrae Regan, 1906
Heterogramma commbae Regan, 1906
Heterogramma corumbae Eigenmann and Ward, 1907.

Original description: Regan, C. T. (1906): A revision of the South-American cichlid genera *Retroculus*, *Geophagus*, *Heterogramma*, and *Biotoecus*. *Ann. Mag. Nat. Hist.*, ser. 7, vol. 17, p. 64–65 (*Heterogramma commmbrae*).

Distribution: River basin of the Rio Paraguai; type specimen near the town of Corumba in the state of Mato Grosso (Brazil).

First importation into Germany: 1906 by the firm Siggelkow (Hamburg).

Characters:
fins: D XV–XVII/5–6, A III–IV/5–7, P 11–12
scales: mLR 22
total length: male up to 60 millimeters; female up to 40 millimeters.

Sex differences: Scarcely observable in younger specimens. Older males exhibit larger and more pointed fins than do females. Brood-tending females are a conspicuous yellow color.

Aquarium keeping: As in *Apistogramma agassizii*.

Spawning: As in *Apistogramma agassizii*.

Apistogramma eunotus
Kullander, 1981

Explanation of the scientific name: Refers to the hump-backed appearance of this species.
eu- (Gk.) = well, true; *notos* (Gk.) = back.

Other name in the literature:
Apistogramma amoenus Regan, 1913.

Original description: Kullander, S. O. (1981): Description of a new species of *Apistogramma* (*Teleostei*: *Cichlidae*) from the upper Amazonas basin. Ergebnisse der Argentinien-Peru-Expedition Dr. K. H. Lüling 1978. *Bonn. Zool. Beitr.*, nr. 32, p. 183–194 (*Apistogramma eunotus*).

Apistogramma caetei.

Apistogramma commbrae.

Apistogramma eunotus.

Apistogramma gephyra.

This is the tank setup in which the author successfully spawned *Apistogramma borelli*. But this setup is especially suitable for any small dwarf cichlid that likes to breed in a cave.

Distribution: River basin of the Rio Ucayali near Pucallpa (Peru), dark water at Campo Verde (according to Lüling, 1981); middle and lower Rio Javari in the region of the upper course of the Amazon (Solimões) at Tabatinga (Peru, Brazil).

Habitat: Close to shore, where riparian growth hangs in the water. Soft water with a *p*H of from 6.3 to 6.4 and an electrical conductivity of 22 *u* S at a water temperature of from 25° to 27° C (after Lüling 1981).

First importation into Germany: 1981 by several members of the Deutschen Cichliden-Gesellschaft.

Characters:
fins: D XIV-XV/6–7, A III/6–7, P 11–12, C 16
scales: mLR 21–23
total length: male up to 80 millimeters; female up to 60 millimeters.

Sex differences: Males have elongated dorsal and anal fins; females have a yellowish-gray ground color.

Aquarium keeping: As in *Apistogramma agassizii*.

Spawning: As in *Apistogramma agassizii*.

Apistogramma elizabethae
Kullander, 1980

Explanation of the scientific name: Refers to Elizabeth Agassiz, the wife of Prof. J. L. R. Agassiz.

Original description: Kullander, S. O. (1980): A taxonomical study of the genus *Apistogramma* Regan, with a revision of Brasilian and Peruvian species (Teleostei: Percoidei: Cichlidae). *Bonn. Zool. Monograph.* nr. 14, p. 103–106 (*Apistogramma elizabethae*).

Distribution: Tributary of the Rio Uaupés at Trovao and in a small tributary of the Lago Penera in the state of Amazonas (Brazil).

First importation into Germany: So far has not yet been imported alive.

Characters:
fins: D XIV-XV/5–7, A III/5–7

total length: male 56.7 millimeters; female 31.9 millimeters (after Kullander, 1980).

Aquarium keeping: As in *Apistogramma agassizii*.

Spawning: As in *Apistogramma agassizii*.

Apistogramma geisleri
Meinken, 1971
Geisler's Dwarf Cichlid

Explanation of the scientific name: Refers to Dr. R. Geisler, the discoverer of this species of fish.

Original description: Meinken, H. (1971): *Apistogramma geisleri* n. sp. und *Apistogramma borellii* (Regan, 1906) aus dem Amazonas-Becken (Pisces: Teleostei: Cichlidae).

Senckenbergiana biol., 52, (1/2), p. 35-40 (*Apistogramma geisleri*).

Distribution: Rio Curucamba at Obidos in the state of Para (Brazil).

First importation into Germany: 1971 by Geisler.

Characters:

fins: D XV/6-7, A III/6–7, P 11–12, V I/6

scales: mLR 22–23

total length: male up to 70 millimeters; female up to 50 millimeters.

Sex differences: Males more brightly colored and with anal and dorsal fins tapering to points. In females these fins are rounded off.

Aquarium keeping: As in *Apistogramma agassizii.*

Spawning: As in *Apistogramma agassizii.*

Apistogramma gephyra
Kullander, 1980

Explanation of the scientific name: Refers to the position of this species between *Apistogramma pertensis* and *Apistogramma agassizii.*

gephyra (Gk.) = bridge.

Original description: Kullander, S. O. (1980): A taxonomical study of the genus *Apistogramma* Regan, with a revision of Brasilian and Peruvian species (Teleostei: Percoidei: Cichlidae). *Bonn. Zool. Monograph.,* nr. 14, p. 131–134 (*Apistogramma gephyra*).

Distribution: Anavilhanas Archipelago on the left bank of the Rio Negro, 100 kilometers above Manaus in the State of Amazonas; in the Lago Jurucui and the Igarapé Grande at Santarem in the State of Para (Brazil).

First importation into Germany: 1982.

Characters:

fins: D XV-XVI/6–8, A III/6–7

total length: male up to 60 millimeters; female up to 50 millimeters.

Sex differences: Males are more colorful and have elongated dorsal and anal fins; females have rounded-off fins and a yellowish coloration.

Aquarium keeping: As in *Apistogramma agassizii.*

Spawning: As in *Apistogramma agassizii.*

Comments: This species is remarkably similar to *Apistogramma agassizii.*

Apistogramma gibbiceps
Meinken, 1969
Black-Banded Dwarf Cichlid

Explanation of the scientific name: Refers to the humped head (forehead protuberance) exhibited by the preserved males in the determination, but which is not observed in living specimens.

gibbus (L.) = gibbous; *kephale* (Gk.) = head.

Original description: Meinken, H. (1969): *Apistogramma gibbiceps* n. sp. aus Brasilien (Pisces: Teleostei: Cichlidae). *Senckenbergiana biol., 50,* p. 91–96 (*Apistogramma gibbiceps*).

Distribution: Apparently the Rio Negro region in the state of Amazonas (Brazil).

First importation into Germany: 1972.

Characters:

fins: D XIV-XV/6–7, A III/6–7

scales: mLR 23

total length: male up to 80 millimeters; female up to 60 millimeters.

Sex differences: Males are more colorful and in some cases have a yellow head. Their tail fin has two points. Females are yellow-gray with a rounded-off tail fin.

Aquarium keeping: As in *Apistogramma agassizii.*

Spawning: As in *Apistogramma agassizii.*

Apistogramma gossei Kullander, 1982
Gosse's Dwarf Cichlid

Explanation of the scientific name: Refers to the discoverer of this species, the Belgian Dr. J. P. Gosse.

Original description: Kullander, S. O. (1982): Description of a new species of *Apistogramma* Regan, from the Oyapock and Approuague River systems (Teleostei: Cichlidae). *Cybium, 6* (4), p. 65–72 (*Apistogramma gossei*).

Distribution: River systems of the Rio Approuague and the Rio Oyapock in French Guiana and bordering Brazil.

First importation into Germany: This species has not yet been imported alive.

Characters:

fins: D XV-XVI/6–7, A III/5–7, P 11–16

scales: mLR 22–23

total length: male up to 60 millimeters; female up to 40 millimeters.

The male *Apistogramma gibbiceps* pursues a suitable female. ▲

The female follows the male into his cave. ▲

The female *gibbiceps* starts spawning. ▲

She lays her eggs on the top of the coconut shell cave. ▲

The pair guard the nest. ▲

The female guarding the spawning site. ▼

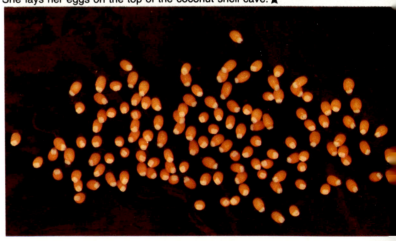

Above & below: Closeups of the spawn of *Apistogramma gibbiceps*..

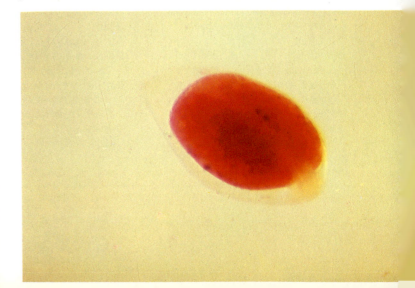

A male *Apistogramma hongsloi.* ▲

A female *Apistogramma hongsloi.* ▲

The female deposits her spawn on the ceiling of her cave. ▲

When she leaves the spawning site, the male enters to fertilize the eggs. ▲

If the male doesn't leave soon enough, the female returns and lays more eggs. ▲

The male waits intently to fertilize the new spawn. ▲

The female stops spawning for a while, but does not allow the male to fertilize the few new eggs she laid.

Then the female commences spawning once again. ▼

Apistogramma gibbiceps, color variety.

Apistogramma gibbiceps, color variety.

Apistogramma gibbiceps, color variety.

Apistogramma hippolytae.

Apistogramma hippolytae ♂, courtship coloration.

Apistogramma hippolytae
Kullander, 1982
Two-Spot Dwarf Cichlid

Explanation of the scientific name: Refers to the legendary Greek figure Hippolyta, the queen of the Amazons.

Other name in the literature: *Apistogramma ambloplitoides* Goldstein, 1973.

Original description: Kullander, S. O. (1982): Beschreibung einer neuen *Apistogramma*-Art aus Zentral-Amazonien (Teleostei: Cichlidae). *DCG-Informationen*, 13. Jg., H. 10, p. 181–193 (*Apistogramma hippolytae*).

Distribution: River system of the Rio Solimões (Igarapé of the Lago Manacapuru), central lake on Ilha des Buiu-acu, near the Rio Urubaxi in the river system of the Rio Negro in the State of Amazonas (Brazil).

First importation into Germany: 1982.

Characters:
fins: D XIV-XV/6–7, A III/6–7, P 12–16
total length: male up to 70 millimeters; female up to 50 millimeters.

Aquarium keeping: As in *Apistogramma agassizii*.

Spawning: As in *Apistogramma agassizii*.

Apistogramma hoignei
Meinken, 1965
Hoigne's Dwarf Cichlid

Explanation of the scientific name: Refers to L. Hoigne, the discoverer of this species.

Original description: Meinken, H. (1965): Eine neue *Apistogramma*-Art aus Venezuela (Pisces: Percoidea: Cichlidae). *Senckenbergiana biol.* 46(4), p. 257–263 (*Apistogramma hoignei*).

Distribution: Tributaries of the marshes on the lower course of the Rio Portuguesa west of the villages of Sta. Rosa and Camaguan, on the highway from Calabozo on the southern end of the lake-like enlargement 'Embalse del Guárico' of the Rio Guárico in the direction of San Fernando on the middle course of the Rio Apure, in the state of Guárico, Venezuela. Additional collecting sites: River basins of the Rio Orinoco, Rio Vichada, and Rio Meta in Colombia.

Characters:
fins: D XVI/6–7, A III/6, V I/5, C 14
scales: mLR 23–24
total length: males probably up to 70 millimeters; females 41.5 millimeters.
The original description is based on two females. The largest female (type specimen) had a length of 41.5 millimeters.

Aquarium keeping: As in *Apistogramma agassizii*.

Spawning: As in *Apistogramma agassizii*.

Apistogramma hongsloi
Kullander, 1979
Hongslo's Dwarf Cichlid

Explanation of the scientific name: Refers to the discoverer of this species, Th. Hongsloi, who also submitted *Apistogramma iniridae*, *Apistogramma macmasteri*, and *Apistogramma viejita* for determination.

Original description: Kullander, S. O. (1979): Species of *Apistogramma* (Teleostei: Cichlidae) from the Orinoco drainage basin, South America,

with descriptions of four new species. *Zool. Scripta, Umea, Sweden*, vol. 8, p. 69–79 (*Apistogramma hongsloi*).

Distribution: According to Kullander (1979), in Colombia at the mouth of the Rio Guarrojo in the Rio Vichada as well as about 155 kilometers northeast of there, on the Caño Perro, a tributary of the Rio Meta.

Habitat: The Caño-Perro collection was caught in a savanna stream that was bordered by a 25-meter-wide strip of forest. The water was clear, colorless, and slow-flowing. pH 4–4.5. Temperature 27–29° C (air) and 26° C (water). The type locality is a savanna lagoon, in which the water was cloudy and a beige color at the time of collecting. pH 5.5. Temperature 29.5° C (air) and 31° C (water, sunny). No other *Apistogramma* species were caught at the *Apistogramma hongsloi* locality. (Kullander, 1979.)

First importation into Germany: 1975 by Dr. Fröhlich (Lubeck).

Characters:
fins: D XIV-XV/6–8, A III/6
scales: mLR 21–23
total length: male up to 80 millimeters; female up to 50 millimeters.

Apistogramma hoignei.

Apistogramma hongsloi.

Apistogramma hongsloi, spawning.

Apistogramma hongsloi, ♀ under the clutch.

Apistogramma hongsloi, color variety.

Sex differences: Older males exhibit a red spot at the base of the tail fin that continues as narrow stripes along the base of the anal fin up to the area of the anus. Males have elongated fins. Females are yellowish with a dark longitudinal stripe. In breeding coloration only seven dark lateral spots are visible in place of the longitudinal stripe.

Aquarium keeping: As in *Apistogramma agassizii.*

Spawning: As in *Apistogramma agassizii.*

Apistogramma inconspicua
Kullander, 1982

Explanation of the scientific name: Refers to the barely present belly stripe, a character used to distinguish this species from *Apistogramma commbrae*, in which the belly stripe is very conspicuous.

inconspicuous (Eng.) = barely present, not striking.

Original description: Kullander, S. O. (1983): Cichlid

fishes from the La Plata basin. Part IV. Review of the *Apistogramma* species (Teleostei: Cichlidae). *Zool. Scripta, Umea, Sweden,* vol. 11, nr. 4, p. 307–310 (*Apistogramma conspicua*).

Distribution: River basin of the Rio Paraguai in Bolivia.

First importation into Germany: 1980

Characters:
fins: D XV/6–7, A III/6, P 11–12
scales: mLR 22
total length: male up to 70

millimeters; female up to 40 millimeters.

Aquarium keeping: As in *Apistogramma agassizii.*

Spawning: As in *Apistogramma agassizii.*

Apistogramma iniridae
Kullander, 1979
Threadfin Dwarf Cichlid

Explanation of the scientific name: Refers to the Rio Inirida, the main river of the region, in which Th. Hongslo caught this fish.

Original description: Kullander, S. O. (1979): Species of *Apistogramma* (Teleostei: Cichlidae) from the Orinoco drainage basin, South America, with descriptions of four new species. *Zool. Scripta, Umea, Sweden,* vol. 8, p. 69–79 (*Apistogramma iniridae*).

Distribution: After Kullander (1979): So far in Puerto Inirida (Obando) and in the Caño Bocon, not far from the mouth, possibly also near San Fernando de Atabapo at the mouth of the Rio Atabapo (probable provenance of the Warrin specimens) (Colombia).

Habitat: Standing bodies of flood water and flowing water of tea-brown color.

First importation into Germany: 1978.

Characters:
fins: D XIV-XV/6–8, A III/6–7
scales: mLR 22–23

total length: male up to 70 millimeters; female up to 60 millimeters.

Sex differences: Males have larger fins and more conspicuous markings. Females are smaller and yellowish colored.

Aquarium keeping: As in *Apistogramma agassizii*, although at least wild-caught fish and the first generations of offspring need soft water with a *p*H of between 5 and 6. It is therefore appropriate to allow the water to flow through a peat filter for some time at each water change.

Spawning: Here one should use a tank without plants; nevertheless, one should furnish it with numerous resinous bog roots, in order to offer sufficient hiding places. Based on my own experiences, the successful spawning of this species is possible only with extremely soft water at a *p*H of about 5. Better than the acidification of the water with diluted phosphoric acid (caution with the dosage) is the constant filtration of the water over peat. In contrast to the majority of *Apistogramma* species, after spawning the male continues to stay in the immediate vicinity of the spawning cavity.

Apistogramma linkei
Koslowski, 1985
Linke's Dwarf Cichlid

Explanation of the scientific name: Refers to H. Linke (West Berlin), who collected most of the type material.

Other name in the literature:
Apistogramma amoenus Lüling, 1969.

Original description:
Koslowski, I. (1985): Descriptions of new species of *Apistogramma* (Teleostei: Cichlidae) from the Rio Mamoré system in Bolivia. *Bonn. Zool. Beitr.,* 36, p. 151–161 (*Apistogramma linkei*).

Distribution: In Bolivia . . . "known so far from the Rio Mamoré system, from Santa Cruz in the South to Trinidad in the North and from the nearby Rio Surucusi, a tributary of the Rio San Miguel." (Koslowski, 1985).

First importation into Germany: 1969 by K. H. Lüling.

Characters:
fins: D XV-XVII/6–7, A III/6–7, P 11–13

scales: mLR 22–23

total length: male up to 60 millimeters; female up to 40 millimeters.

Sex differences: Males have more conspicuous markings and elongated fins.

Aquarium keeping: As in *Apistogramma agassizii.*

Spawning: As in *Apistogramma agassizii.*

Apistogramma luelingi
Kullander, 1976
Lüling's Dwarf Cichlid

Explanation of the scientific name: Refers to Dr. K. H. Lüling, who collected this species in 1966.

Other name in the literature:
Apistogramma borellii Meinken, 1969.

Original description:
Kullander, S. O. (1976): *Apistogramma luelingi* spec. nov., a new Cichlid fish from Bolivia (Teleostei: Cichlidae). *Bonn. Zool. Beitr.,* 27, p. 258–266 (*Apistogramma luelingi*).

Distribution: Vicinity of Todos Santos (near the Rio Chaparé) in Bolivia.

Habitat: Rain forest river with gray-brown water; *p*H about 6; water temperature during day about 26° C.

First importation into Germany: 1966 by Dr. K. H. Lüling (Bonn).

Characters:
fins: D XV-XVI/5–6, A III-IV/5–6, P 11–14, V I/5, C 16

scales: mLR 19–23

total length: male up to 70 millimeters; female up to 40 millimeters.

Aquarium keeping: As in *Apistogramma agassizii.*

Spawning: As in *Apistogramma agassizii.*

Apistogramma macmasteri
Kullander, 1979
Villavicencio Dwarf Cichlid

Explanation of the scientific name: Refers to M. McMaster, who brought this fish to Kullander's attention in 1973.

Other names in the literature:
Apistogramma taeniata Eigenmann, 1922
Apistogramma commbrae Eigenmann, 1922
Apistogramma taeniatus Fowler, 1942
Apistogramma ornatipinnis Staeck, 1974.

Original description:
Kullander, S. O. (1979): Species of *Apistogramma* (Teleostei: Cichlidae) from the Orinoco drainage basin, South America, with descriptions of four new species. *Zool. Scripta, Umea, Sweden,* vol. 8, p. 68–79 (*Apistogramma macmasteri*).

Distribution: Streams in the drainage area of the Rio Guatiquia near the town of Villavicencio, as well as in the Rio Metica at Puerto López in the Departemento Meto (Colombia).

Habitat: Clear-water streams with a total hardness of under 1° dH and a *p*H of between 5.5 and 6; water temperature 27° C. Leaf-covered substrate, in part with aquatic plant growth, water depth between 10 and 15 centimeters. The fish principally stayed in places in the stream where they were hidden from view.

First importation into Germany: 1969.

Characters:
fins: D XV-XVI/6–8, A III/6–7

scales: mLR 22–23

total length: male up to 100 millimeters; female up to 60 millimeters.

Sex differences: Males are more colorful with fins that taper to a point. The red-colored margins of the tail fin are characteristic. Females are yellowish with dark spots.

Aquarium keeping: As in *Apistogramma agassizii.*

Spawning: As in *Apistogramma agassizii.*

Apistogramma meinkeni
Kullander, 1980
Meinken's Dwarf Cichlid

Explanation of the scientific name: Refers to Dr. H. Meinken,

Apistogramma inconspicua, courtship coloration.

Apistogramma inconspicua, normal coloration.

Apistogramma iniridae.

Apistogramma linkei.

Apistogramma luelingi.

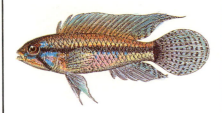

who devoted his attention particularly to the genus *Apistogramma*.

Original description: Kullander, S. O. (1980): A taxonomical study of the genus *Apistogramma* Regan, with a revision of the Brasilian and Peruvian species (Teleostei: Percoidei: Cichlidae). *Bonn. Zool. Monograph.,* nr. 14, p. 118–122 (*Apistogramma meinkeni*).

Distribution: Tributary of the Rio Uaupés at Travao in the state of Amazonas (Brazil).

First importation into Germany: 1982.

Characters:
fins: D XV/6–7, A III/6–7
scales: mLR 22–24
total length: male up to 60 millimeters; female up to 45 millimeters.

Sex differences: Males have fins tapered to a point and are more colorful, especially in the head area. Females are yellowish.

Aquarium keeping: As in *Apistogramma agassizii.*

Spawning: As in *Apistogramma agassizii.*

Apistogramma moae
Kullander, 1980
Moa Dwarf Cichlid

Explanation of the scientific name: Refers to the Rio Moá, in which the species was collected.

Other name in the literature: *Apistogramma amoenus* Regan, 1913.

Original description: Kullander, S. O. (1980): A taxonomical study of the genus *Apistogramma* Regan, with a revision of Brasilian and Peruvian species (Teleostei: Percoidei: Cichlidae). *Bonn. Zool. Monograph.,* nr. 14, p. 61–65 (*Apistogramma moae*).

Distribution: Igarapé Sao Salvador, tributary of the Rio Moá near Cruzeiro do Sul. The Rio Moá is a tributary of the Rio Jurua near the Peruvian border of the State of Acre (Brazil).

First importation into Germany: 1982.

Characters:
fins: D XV/7, A III/7

scales: mLR 23
total length: male up to 60 millimeters; female up to 40 millimeters.

Sex differences: Males have fins that taper to points; females are yellowish with rounded-off fins.

Aquarium keeping: As in *Apistogramma agassizii.*

Spawning: As in *Apistogramma agassizii.*

Apistogramma nijsseni
Kullander, 1979
Nijssen's Dwarf Cichlid

Explanation of the scientific name: Refers to Dr. H. Nijssen, the curator of the Amsterdam Museum of Zoology.

Original description: Kullander, S. O. (1979): Description of a new species of the genus *Apistogramma*

Apistogramma meinkeni.

112

Apistogramma macmasteri.

Apistogramma macmasteri, color variety.

Apistogramma macmasteri, color variety.

Apistogramma nijsseni, young ♂.

Apistogramma nijsseni, adult ♂.

Apistogramma nijsseni, ♀ in brood-care coloration on the clutch.

(Teleostei: Perciformes: Cichlidae) from Peru. *Revue Suisse Zool.* vol. 86(4), p. 937–945 (*Apistogramma nijsseni*).

Distribution: Lower Rio Ucayali, 14 kilometers from Jenaro Herrara in the direction of Colonia Angamos in the Departemento Loreto (Peru).

Habitat: Small, slow-flowing water courses with dark-brown, clear water; the bottom covered with leaves and branches. Total hardness under 1° dH, *p*H about 5.5, water temperature about 27° C.

First importation into Germany: 1978 by Patrik de Rham (Lausanne).

Characters:
fins: D XV-XVI/7–8, A III/7
scales: mLR 22
total length: male up to 80 millimeters; female up to 60 millimeters.

Sex differences: Males are bluish; the tail fin has a distinct reddish edging. Females are yellow with large dark-green, during brood care almost black, spots in the middle of the body, on the lower gill cover, and on the caudal peduncle; the tail fin also has a reddish edging.

Aquarium keeping: As in *Apistogramma agassizii*.

Spawning: As in *Apistogramma agassizii*, although water with

almost no hardness absolutely should be used that has been brought to a *p*H of below 6 by means of filtration through peat.

Comments: *Apistogramma nijsseni* females differ strikingly in coloration from all previously known *Apistogramma* species (see color plates).

Apistogramma ortmanni
(Eigenmann, 1912)
Ortmann's Dwarf Cichlid

Explanation of the scientific name: Refers to the proper name Ortmann (no details in the original description).

Apistogramma ortmanni.

Apistogramma pleurotaenia.

Apistogramma regani, normal coloration.

Apistogramma regani, courtship coloration.

Apistogramma staecki.

Other name in the literature:
Heterogramma ortmanni Eigenmann, 1912.

Original description: Eigenmann, C. H. (1912): The freshwater fishes of British Guiana, including a study of the ecological groupings of species, and the relation of the fauna of the plateau to that of the lowlands. *Mem. Carnegie Mus.*, vol. 5, p. 506–507 (*Heterogramma ortmanni*).

Distribution: Packeoo Fall, Erukin, Kangaruma Gluck Island, Konawaruk, Rockstone; river basin of the Potara River and the Rupununi River (tributary of the Essequibo River in Guayana).

First importation into Germany: 1934 by Aquarium Hamburg.

Characters:
fins: D XV-XVI/7, A III/6–7
scales: mLR 22–24
total length: male up to 80 millimeters; female up to 60 millimeters.

Sex differences: Dorsal and anal fins of males taper to a point; females have rounded-off fins.

Aquarium keeping: As in *Apistogramma agassizii.*

Spawning: As in *Apistogramma agassizii.*

Apistogramma parva
Ahl, 1931

Explanation of the scientific name: Refers to the small size of the type specimen.
parvus (L.) = small.

Original description: Ahl, E. (1931): Neue Süsswasserfische aus dem Stromgebiet des Amazonasstromes. *Sber. Ges. naturf. Freunde Berlin*, p. 210–211 (*Apistogramma parva*).

Distribution: Rio Capim in the state of Para (Brazil).

First importation into Germany: Not yet imported alive.

Characters:
fins: D XV/6, A III/6
scales: mLR 24
total length: type specimen 22 millimeters (sex no longer determinable).

Comments: This species is not completely certain, since Ahl's description does not agree fully with the type specimen. For a definite determination additional material is needed. The collecting region of the type specimen is, however, not precisely located.

Apistogramma personata
Kullander, 1980

Explanation of the scientific name: Refers to the dark band between the eyes.
personatus (L.) = masked.

Original description: Kullander, S. O. (1980): A taxonomical study of the genus *Apistogramma* Regan, with a revision of Brasilian and Peruvian species (Teleostei: Percoidei: Cichlidae). *Bonner Zool. Monograph.*, nr. 14, p. 111–114 (*Apistogramma personata*).

Distribution: Rio Uaupés at Assai in the state of Amazonas (Brazil).

First importation into Germany: 1981.

Characters:
fins: D XV/6–8, A III/5–7
scales: mLR 22–24
total length: male up to 70 millimeters; female up to 50 millimeters.

Sex differences: Males with fins tapered to a point and double-pointed tail fin; females have rounded-off fins.

Aquarium keeping: As in *Apistogramma agassizii.*

Spawning: As in *Apistogramma agassizii.*

Apistogramma pertensis
(Haseman, 1911)
Amazonas Dwarf Cichlid

Explanation of the scientific name: Apparently refers to the former affiliation with *Apistogramma taeniata.*
pertinere (L.) = belonging to.

Other names in the literature:
Apistogramma taeniatum pertense Haseman, 1911
Apistogramma taeniatum pertense Fowler, 1954.

Original description: Haseman, J. D. (1911): An annotated catalog of the Cichlid fishes collected by the expedition of the Carnegie Museum to Central South America, 1907–10. *Ann. Carnegie Mus.*, 7, p. 359 (*Heterogramma taeniatum pertense*).

Distribution: River basin of the Amazon from Santarem right into the upper Rio Negro River basin (Brazil).

First importation into Germany: 1980.

Apistogramma pertensis.

Characters:
fins: D XIV-XVI/6–8, A III/4–7
scales: mLR 22-24
total length: male up to 70 millimeters; female up to 50 millimeters.

Sex differences: Males have strikingly large fins, with the ventral fins greatly elongated.

Aquarium keeping: As in *Apistogramma agassizii.*

Spawning: As in *Apistogramma agassizii.*

Apistogramma piauiensis
Kullander, 1980
Piaui Dwarf Cichlid

Explanation of the scientific name: Refers to the State of Piaui in Brazil, where the species was found.

Original description: Kullander, S. O. (1980): A taxonomical study of the genus *Apistogramma* Regan, with a revision of Brasilian and Peruvian species (Teleostei: Percoidei: Cichlidae). *Bonner Zool. Monograph.*, nr. 14, p. 79–82 (*Apistogramma piauiensis*).

Distribution: According to Kullander (1980), the delta of the Rio Longa in the Rio Parnaiba and the Lagoa Seca on the Rio Parnaiba in the Brazilian state of Piaui.

Habitat: The Lagoa Seca (Dry Lake) is located 1 or 2 kilometers from the main stream of the Rio Parnaiba and Barra do Longa, and is located in the flood plain of the Rio Parnaiba. At the time of capture it was approximately 20 to 400 meters wide and 1 kilometer long with a maximum depth of 1 meter. The bottom was mostly muddy, with rocks in some places.

Virtually no vegetation of large size was found there. At higher water levels submerged vegetation, including grasses and bushes, as well as transitions to marshy areas with aquatic-plant vegetation could be present. With *Apistogramma piauiensis* about 20 to 30 fish species, including many typical of the main stream, were found (T. R. Roberts, in litt.). (Kullander, 1980.)

First importation into Germany: This species apparently has not yet been imported alive.

Characters:
fins: D XV-XVI/6–7, A III/6
scales: mLR 23
total length: female 30.6 millimeters (type specimen after Kullander 1980).

Comments: The determination was made on the basis of a female, so that no data exist on the appearance of the male.

Apistogramma pleurotaenia
(Regan, 1909)
Chequered Dwarf Cichlid

Explanation of the scientific name: Refers to the dark lateral band.
pleura (L.) = side; *taenia* (L.) = band, stripe.

Other names in the literature:
Geophagus taeniatus Thumm, 1906
Heterogramma pleurotaenia Regan, 1909
Heterogramma borelli Haseman, 1911.

Original description: Regan, C. T. (1909): Description of a new Cichlid fish of the genus *Heterogramma* from the La Plata. *Ann. Mag. Nat. Hist.,* ser. 8, vol. 3, p. 270 (*Heterogramma pleurotaenia*).

Distribution: La Plata region in southern Brazil (type specimen).

First importation into Germany: 1905 by Kittler (Hamburg).

Characters:
fins: D XVI/6, A IV/5 (apparently A III/6 is correct)
scales: mLR 23
total length: male up to 75 millimeters; female up to 40 millimeters (type specimen after Regan 1909).

Sex differences: Dorsal and anal fins of males taper to a point; females have rounded-off fins.

Aquarium keeping: As in *Apistogramma agassizii.*

Spawning: As in *Apistogramma agassizii.*

Apistogramma pulchra
Kullander, 1980

Explanation of the scientific name: Refers to the beautiful coloration and slender build.
pulcher (L.) = beautiful.

Original description: Kullander, S. O. (1980): A taxonomical study of the genus *Apistogramma* Regan, with a revision of Brasilian and Peruvian species (Teleostei: Percoidei: Cichlidae). *Bonner Zool. Monograph.*, nr. 14, p. 135–138 (*Apistogramma pulcher*).

Distribution: Rio Preto, a small tributary of the Rio Candeias (tributary of the Rio Madeira) near Porto Velho in the State of Rondonia (Brazil).

First importation into Germany: Apparently not yet imported alive.

Characters:
fins: D XIV-XVI/7–8, A III/6–7
scales: mLR 23
total length: largest male 43.9 millimeters; largest female 26.8 millimeters (type specimens after Kullander 1980).

Sex differences: Fins of males taper to a point; females have rounded-off fins.

Aquarium keeping: As in *Apistogramma agassizii.*

Spawning: As in *Apistogramma agassizii.*

Apistogramma regani
Kullander, 1980
Regan's Dwarf Cichlid

Explanation of the scientific name: Refers to the British ichthyologist C. T. Regan.

Other names in the literature:
Heterogramma ortmanni Haseman, 1911
Heterogramma taeniatum Miranda-Ribeiro, 1918
Apistogramma taeniatum Marlier, 1967
Apistogramma borellii Meinken, 1971.

Original description: Kullander, S. O. (1980): A taxonomical study of the genus *Apistogramma* Regan, with a revision of Brasilian and Peruvian species (Teleostei: Percoidei: Cichlidae). *Bonner Zool. Monograph.*, nr. 14, p. 65–72 (*Apistogramma regani*).

Distribution: Region of the Anavilhanas Peninsula above Manaus and Lago Redondo on the right bank of the Amazon, 25 kilometers southwest of Manaus in the State of Amazonas (Brazil).

First importation into Germany: 1914.

Characters:
fins: D XIV-XVI/6–8, A III/6–7
scales: mLR 22–23
total length: male up to 70 millimeters; female up to 50 millimeters.

Sex differences: Dorsal and anal fins of males taper to a point; females have rounded-off fins.

Aquarium keeping: As in *Apistogramma agassizii.*

Spawning: As in *Apistogramma agassizii.*

Apistogramma roraimae
Kullander, 1980
Roraima Dwarf Cichlid

Explanation of the scientific name: Refers to the Brazilian State of Roraima, in which the area of occurrence is located.

Original description: Kullander, S. O. (1980): A taxonomical study of the genus *Apistogramma* Regan, with a revision of Brasilian and Peruvian species (Teleostei: Percoidei: Cichlidae). *Bonner Zool. Monograph.*, nr. 14, p. 138–141 (*Apistogramma roraimae*).

Distribution: Tributaries of the Rio Branco on the road from Boa Vista to Caracarai in the state of Roraima (Brazil).

Characters:
fins: D XV/7, A III/6–8
scales: mLR 22–23
total length: male 32.6 millimeters; female 27.8 millimeters (type specimens after Kullander 1980).

Apistogramma staecki
Koslowski, 1985
Staecks Dwarf Cichlid

Explanation of the scientific name: Refers to Dr. W. Staeck, one of the collecters of the type material.

Original description: Koslowski, I. (1985): Descriptions of new species of *Apistogramma* (Teleostei: Cichlidae) from the Rio Mamoré system in Bolivia. *Bonn. Zool. Beitr.*, 36, p. 145–162 (*Apistogramma staecki*).

Distribution: Rio Mamoré River system south of the town of Trinidad (Bolivia), between Guajara Mirim and Mato Grosso (Brazil).

Habitat: Very shallow lagoons among dead leaves and branches.

First importation into Germany: 1983 by Linke and Staeck (West Berlin).

Characters:
fins: D XV-XVI/5–6, A III/6–7, P 11–12
scales: mLR 22-23
total length: male up to 60 millimeters; female up to 40 millimeters.

Sex differences: Males more colorful, with dorsal and anal fins that taper to a point; females yellowish with rounded-off fins.

Aquarium keeping: As in *Apistogramma agassizii.*

Spawning: As in *Apistogramma agassizii.*

Apistogramma steindachneri.

Apistogramma steindachneri, courting ♂.

Apistogramma steindachneri, ♀ spawning.

Apistogramma steindachneri, ♂ fertilizing.

Apistogramma steindachneri, ♀ guards the clutch.

Spawning series of *Apistogramma steindachneri.*

Apistogramma steindachneri
(Regan, 1908)
Steindachner's Dwarf Cichlid

Explanation of the scientific name: Refers to F. Steindachner, who also described *Apistogramma agassizii*, among others.

Other names in the literature:
Heterogramma steindachneri Regan, 1908
Apistogramma ortmanni rupununi Fowler, 1914
Apistogramma ornatipinnis Ahl, 1936
Apistogramma wickleri Meinken, 1960.

Original description: Regan, C. T. (1908): Description of a new Cichlid fish of the genus *Heterogramma* from Demerara. *Ann. Mag. Nat. Hist.,* ser. 8, vol. 1, p. 370–371 (*Heterogramma steindachneri*).
Distribution: The river basins of the Essequibo River and the Demerara River (Guyana).
First importation into Germany: 1936 by Fritz Mayer (Hamburg).

Characters:
fins: D XV/7, A III/6
scales: mLR 24
total length: male up to 100 millimeters; female up to 70 millimeters.
Sex differences: Males have large dorsal and anal fins that taper to a point; the tail fin has two points. The coloration is more conspicuous. Females have rounded-off fins.
Aquarium keeping: As in *Apistogramma agassizii.*
Spawning: As in *Apistogramma agassizii.*

Apistogramma sweglesi
Meinken, 1961
Swegle's Dwarf Cichlid

Explanation of the scientific name: Refers to the collector of this species, the American Kyle Swegles.
Original description: Meinken, H. (1961): Drei neu eingefuhrte *Apistogramma*-Arten aus Peru, eine davon wissenschaftlich neu. *Die Aquar.- u. Terr.-Zschr. (DATZ),* 14(5), p. 135–139 (*Apistogramma sweglesi*).
Distribution: According to Meinken (1961) or Swegles, at Letitia (Peru). However, in all probability what is meant is Leticia on the Amazon in Colombia (the border region between Brazil, Colombia, and Peru).

A pair of *Apistogramma steindachneri* spawning. Note the intense red color of their eggs.

First importation into Germany: 1960 by Dr. Schmidt-Focke (Bad Homburg).

Characters:
fins: D XV/7, A III/6, P 11
scales: mLR 24
total length: male 73 millimeters; female 50 millimeters (type specimens after Meinken, 1961).

Sex differences: Males are more conspicuously colored and the tail fin has two points. Females are yellowish with a rounded-off tail fin.

Aquarium keeping: As in *Apistogramma agassizii*.

Spawning: As in *Apistogramma agassizii*.

Comments: This species is doubtful. The type specimens can no longer be traced, so that a revue of Meinken's determination is impossible at present. It is possibly the same as *Apistogramma bitaeniata*.

Apistogramma taeniata
(Günther, 1862)
Banded Dwarf Cichlid

Explanation of the scientific name: Refers to the stripes of the fish.
taenia (L.) = stripe.

Other names in the literature:
Mesops taeniatus Günther, 1862
Geophagus taeniatus Steindachner, 1875
Heterogramma taeniatum Regan, 1906.

Original description: Günther, A. C. L. G. (1862): Catalogue of the fishes in the British Museum. British Mus. London, vol. 4, p. 312 (*Mesops taeniatus*).

Distribution: Rio Cupari (tributary of the Rio Tapajos) about 200 kilometers south of the town of Santarem in the Brazilian state of Para; according to Günther (1862) 800 miles from the ocean.

Characters:
fins: D XV/6, A III/6, P 12
scales: mLR 23
total length: type specimen 42.1 millimeters.

Comments: The description of this fish by Günther (1862) is based on a single specimen. Today the very badly preserved fish scarcely allows a new description. So far fish of this species apparently have not been imported again.

Apistogramma trifasciata
(Eigenmann and Kennedy, 1903)
Three-Striped Dwarf Cichlid

Explanation of the scientific name: Refers to the three dark stripes on the body (lateral band, cheek stripe, and a stripe from the eye in the direction of the anus).
tres (L.) = three; *tri-* (L.) = three- (in compound words); *fasciatus* (L.) = striped.

Other names in the literature:
Biotodoma trifasciatus Eigenmann and Kennedy, 1903
Heterogramma trifasciatum Regan, 1906
Apistogramma trifasciata macilliensis Haseman, 1911
Apistogramma trifasciata haraldschultzi Meinken, 1960.

Original description: Eigenmann, C. H., and Kennedy, C. H. (1903): On a collection of fishes from Paraguay, with a synopsis of the American genera of Cichlids. *Proc. Acad. Nat. Sci.,*

Philadelphia, nr. 56, p. 536 (*Biotodoma trifasciatus*).

Distribution: River basins of the upper Rio Paraguai (Brazil) and the Rio Guapore (near San Antonio).

First importation into Germany: 1959.

Characters:
fins: D XV/6, A III/5
scales: mLR 22
total length: male up to 60 millimeters; female up to 40 millimeters.

Sex differences: Males are more colorful and have elongated rays in the dorsal fin as well as very long ventral fins. Females are yellowish with rounded-off fins.

Aquarium keeping: As in *Apistogramma agassizii*.

Spawning: As in *Apistogramma agassizii*.

Comments: Two subspecies of *Apistogramma trifasciata* were described (*Apistogramma trifasciata macilliensis* and *Apistogramma trifasciata haraldschultzi*). It turned out, however, that in all specimens it was a question of *Apistogramma trifasciata* that vary, in coloration in particular, only because of its large area of occurrence.

Apistogramma uaupesi
Kullander, 1980
Uaupés Dwarf Cichlid

Explanation of the scientific name: Refers to the Rio Uaupés, the region in which the type specimens were collected.

Original description: Kullander, S. O. (1980): A

Apistogramma trifasciata, color variety.

Apistogramma trifasciata, color variety.

Apistogramma trifasciata, color variety.

taxonomical study of the genus *Apistogramma* Regan, with a revision of Brasilian and Peruvian species (Teleostei: Percoidei: Cichlidae). *Bonner Zool. Monograph.*, nr. 14, p. 122-126 (*Apistogramma uaupese*).

Distribution: Stream on the right bank of the Rio Uaupés near the village of Travao in the state of Amazonas (Brazil).

First importation into Germany: Apparently not yet imported alive.

Characters:
fins: D XV/6–7, A III/4–6
scales: mLR 22–24
total length: male 37 millimeters; female 34 millimeters (type specimens after Kullander, 1980).

Apistogramma viejita
Kullander, 1979
Black-Throated Dwarf Cichlid

Explanation of the scientific name: Refers to the native name "vieja" for various *Apistogramma* species in eastern Colombia.

Original description: Kullander, S. O. (1979): Species of *Apistogramma* (Teleostei: Cichlidae) from the Orinoco drainage basin, South America, with descriptions of four new species. *Zool. Scripta, Umea, Sweden,* vol. 8, p. 69–79 (*Apistogramma viejita*).

Distribution: The locality of the type specimen is a stream flowing into the Rio Yucao (tributary river of the Rio Meta in the Orinoco drainage basin) in eastern Colombia.

Habitat: Clear, brownish water with a total hardness of under 1° dH and a *p*H of about 5; sandy substrate, marginal zones with submerged plants.

First importation into Germany: 1982 by Horst Linke (West Berlin).

Characters:
fins: D XV/7, A III/6
scales: mLR 23
total length: male up to 70 millimeters; female up to 40 millimeters.

Sex differences: Males altogether more colorful, with large fins tapering to a point. The margins of the tail fin are red as in *Apistogramma macmasteri*. Females are yellowish with a deep-black belly stripe, which starts at the throat. It is a typical identifying character of the *Apistogramma viejita* female, specifically during the time of brood care.

Aquarium keeping: As in *Apistogramma agassizii.*

Spawning: As in *Apistogramma agassizii.*

Comments: Various color varieties of *Apistogramma viejita* are known. Males strongly resemble those of *Apistogramma macmasteri*, but the dark lateral band is clearly interrupted in *Apistogramma macmasteri*, whereas in *Apistogramma viejita* it usually appears continuous.

Up to now undescribed *Apistogramma* species

On account of the inaccessibility of the habitats and the large distances to the centers of exotic fish collecting areas or fairly large settlements, to this day only a portion of the actually existing species have been regularly imported. With the increasing possibility of pushing on into inaccessible regions, in recent years more and more previously undescribed species have been collected. In this way a large number of new *Apistogramma* species were discovered and imported in the last decade, which, nevertheless, so far have not been scientifically described for the most diverse reasons. In aquarists' circles these fishes received so-called common names. Some of them do not reflect the actual normal appearance of the fishes. For this reason, these *Apistogramma* species will not be identified by these names here.

Genus Apistogrammoides
Meinken, 1965

Explanation of the scientific name: Refers to the similarity of these fish to those of the genus *Apistogramma.*

eidos (Gk.) = appearance, form; *-oides* (L.) = similar, -like.

Original description: Meinken, H. (1965): Uber eine neue Gattung und Art der Familie *Cichlidae* aus Peru (Pisces: Percoides: Cichlidae). *Senckenbergiana biol.* 46(1), p. 47–53.

Distribution: Peru.

Number of species: So far only one species is known.

Generic characters: The forward part of the lateral line over its entire length is always located more than one scale's width from the basis of the dorsal fin. The most important character used to distinguish the member of this genus from those of *Apistogramma* is the differing formation of the anal fin. In *Apistogrammoides* it has eight spines, whereas in *Apistogramma*, as a rule, it exhibits only three spines.

Total length: Up to 50 millimeters.

Type species: *Apistogramma pucallpaensis* Meinken, 1965

Apistogramma species that have not yet been scientifically described.

Characters:

fins: D XVII/6, A VIII/5–6, V I/5–6, P 12–13

scales: mLR 22–23

total length: male up to 50 millimeters; female up to 35 millimeters.

Sex differences: Males are more colorful with larger fins; females are yellow-gray with rounded-off fins.

Aquarium keeping: As in *Apistogramma agassizii*, yet even fairly small tanks are possible. The fish can be kept in them in pairs.

Spawning: As in *Apistogramma agassizii*. The male takes part in brood care to some extent.

Explanation of the scientific name: Refers to the assumption that the species is mouthbrooder. *bios* (Gk.) = life; *oikos* (Gk.) = house, homeland.

Original description: Eigenmann, C. H., and Kennedy, C. H. (1903): On a collection of fishes from Paraguay, with a synopsis of the American genera of Cichlids. *Proc. Acad. Nat. Sci. Phila.*, nr. 56, p. 533.

Distribution: According to Steindachner (1875), Lake Saraca and the tributaries of the Amazon at Villa Bella (Brazil).

Number of species: So far one species is known.

Total length: Up to 100 millimeters.

Type species: *Biotoecus opercularis* (Saraca opercularis Steindachner, 1875).

Apistogrammoides pucallpaensis
Meinken, 1965
Pucallpa Dwarf Cichlid

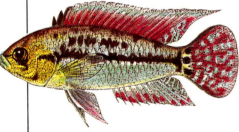

Genus *Biotoecus* Eigenmann and Kennedy, 1903

Biotoecus opercularis
(Steindachner, 1875)

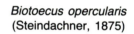

Explanation of the scientific name: Refers to the town of Pucallpa in Peru, near which the fish were caught.

Original description: Meinken, H. (1965): Uber eine neue Gattung und Art der Familie *Cichlidae* aus Peru (Pisces: Percoides: Cichlidae). *Senckenbergiana biol.* 46(1), p. 47–53 (*Apistogrammoides pucallpaensis*).

Distribution: Stream near the town of Pucallpa, which flows into the Rio Ucayali, and in a tributary of the Yarina Cocha (Peru).

First importation into Germany: 1976.

Explanation of the scientific name: Refers to the markings on the gill cover. *operculum* (L.) = covering (gill cover).

Other names in the literature: *Saraca opercularis* Steindachner, 1875.

Original description: Steindachner, F. (1875): Beitrage zur Kenntnis der Chromiden des Amazonenstromes. *Sber. Akad. Wiss., Wien,* 71, p. 125–127 (*Saraca opercularis*).

Distribution: According to Steindachner (1875), Lake Saraca and the tributaries of the Amazon at Villa Bella.

Comments: So far no detailed information on this fish species exists; apparently this species has not yet been imported to Europe.

Apistogrammoides pucallpaensis ♂.

Apistogrammoides pucallpaensis in the act of spawning.

Apistogrammoides pucallpaensis tending the clutch.

Apistogramma species that have not yet been scientifically described.

Genus *Chalinochromis*
Poll, 1974

Explanation of the scientific name: Refers to the dark stripes on the head of the fish and to the former name for cichlids, chromides.
chalinos (Gk.) = rein; *chroma* (Gk.) = color.

Original description: Poll, M. (1974): Contribution à la faune ichthyologique du lac Tanganyika D'après les récoltes de P. Brichard. *Rev. Zool. Bot. Afr.,* vol. 88(1), p. 99–110.

Distribution: Lake Tanganyika (Africa).

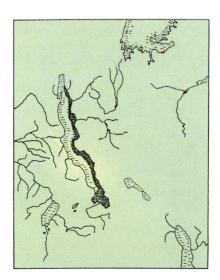

Number of species: So far three species are known, of which one species has been scientifically described.

Generic characters: Six to seven spines in the anal fin; there is a close relationship to the genus *Julidochromis*.

Total length: Up to 120 millimeters.

Type species: *Chalinochromis brichardi* Poll, 1974.

Chalinochromis brichardi Poll, 1974

Explanation of the scientific name: Refers to the Belgian author of books and expert on the fish species of Tanganyika, P. Brichard (Burundi).

Original description: Poll, M. (1974): Contribution à la faune ichthyologique du lac Tanganyika d'après les récoltes de P. Brichard. *Rev. Zool. Bot. Afr.,* vol. 88(1), p. 104–106.

Distribution: Lake Tanganyika.
Habitat: Rocky riparian zones.
First importation into Germany: 1973.
Characters:
fins: D XXII-XXIII/6–7, A VII/5
scales: mLR 35–36
total length: both sexes up to 120 millimeters.

Sex differences: Not externally visible.

Aquarium keeping: For these fish a fairly large aquarium with a capacity of more than 100 liters is suitable, in which they should be kept with other Tanganyika dwarf cichlids. The tank should be furnished with numerous rock structures and aquatic plants, such as *Bolbitis heudelotti*. One should use water that is not too soft, the *p*H of which should not be less than 7. Medium-hard water is best suited for keeping these fish. The water temperature can be between 23° and 28° C.

Spawning: Spawning is not all that easy. The fish need a tank with many cavity-like rock structures. They spawn on the roof of a cave. The clutch and brood are tended only relatively casually. The fry are not kept together in a school by the parents. Nevertheless, they principally stay in the parents' territory. During this time the parents are extremely aggressive toward all other fishes.

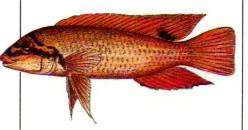

Comments: The other known members of the genus *Chalinochromis* exhibit distinct patterns or markings and up to now have been offered under the names *Chalinochromis* spec. Ndoboi and *Chalinochromis* spec. "bifrenatus" by dealers.

Genus *Julidochromis*
Boulenger, 1898

Explanation of the scientific name: Refers to the similarity to the former marine wrasse group Julidini and to the former name for cichlids, "chromides."
chroma (Gk.) = color.

Original description: Boulenger, G. A. (1898): Report on the collection of fishes made by Mr. J. E. S. Moore in Lake Tanganyika . . . 1895–1896. *Trans. Zool. Soc.,* 15, p. 11.

Distribution: Lake Tanganyika (Africa).

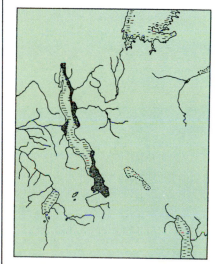

Number of species: So far five species are known.

Generic characters: 20 to 24 spines in the dorsal fin; 7 to 9 spines in the anal fin. The scales are strongly serrated and are very small on the nape.

Total length: 80 to 150 millimeters.

Type species: *Julidochromis ornatus* Boulenger, 1898.

Chalinochromis brichardi.

Chalinochromis spec. Ndoboi.

Chalinohromis spec. "bifrenatus."

Chalinochromis brichardi, portrait.

Julidochromis dickfeldi
Staeck, 1975
Dickfeld's Slender Cichlid

Explanation of the scientific name: Refers to A. Dickfeld, the inspirer of the collecting trip.
Original description: Staeck, W. (1975): A new fish from Lake Tanganyika. *Julidochromis dickfeldi* sp. n. (Pisces: Cichlidae). *Rev. Zool. Bot. Afr.,* Vol. 89, p. 982–986 (*Julidochromis dickfeldi*).
Distribution: Southwestern part of Lake Tanganyika, north of Sumbu National Park in Zambia.
Habitat: Boulder zones and rocky places.
First importation into Germany: 1975 by A. Dickfeld.

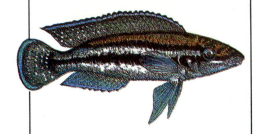

Characters:
fins: D XXIV/5, A VIII–IX/4–5
scales: mLR 35
total length: both sexes up to about 110 millimeters.
Sex differences: Not readily discernible. Upon close examination, one can see an approximately one-millimeter long genital papilla in the male.
Aquarium keeping: As far as possible, tanks more than 80 centimeters long should be used. Numerous rock structures provided with caves and fine-grained gravel as a substrate correspond to the natural ecotope. Medium-hard water with a *p*H of not less than 7 and a water temperature of between 23° and 28° C are favorable. Handling in the tank is disturbing to the fish, since fighting between mates often occurs after it.
Spawning: Spawning is possible in the keeping tank. The eggs are laid on the roof of a cave. As a rule, not more than 40 eggs of a gray-green color are

laid. The larvae hatch at 26° C after about three days and hang on the roof of the cave with the aid of an adhesive gland on the head. About eight days after spawning the fry become free-swimming. The brood care of the parents is cursory.

Julidochromis marlieri Poll, 1956
Checkerboard Slender Cichlid

Explanation of the scientific name: Refers to the Belgian ichthyologist G. Marlier.
Original description: Poll, M. (1956): Poissons Cichlidae. Résultats scientifiques exploration hydrobiologique du Lac Tanganyika 1946–1947. vol. 3, part 5B, p. 1–619 (*Julidochromis marlieri*).
Distribution: Northwestern part of Lake Tanganyika near Makabola and Luhanga in Zaire.
Habitat: Boulder and rock zones.
First importation into Germany: 1958.

Characters:
fins: D XX-XXII/5–6, A VIII-IX/5
scales: mLR 32
total length: male up to 130 millimeters; female up to 120 millimeters.
Sex differences: Scarcely discernible. Males with more pointed genital papilla slanting toward the rear.
Aquarium keeping: As in *Julidochromis dickfeldi*, although, corresponding to the size of the fish, one should not use a tank less than 100 centimeters long.
Spawning: As in *Julidochromis dickfeldi*.
Comments: Depending on the locality, there are different color varieties.

Julidochromis ornatus
Boulenger, 1898
Striped Slender Cichlid

Explanation of the scientific name: Refers to the beautiful coloration of the fish.
ornatus (L.) = adorned.
Original description: Boulenger, G. A. (1898): Report on the collection of fishes made by Mr. J. E. S. Moore in Lake Tanganyika . . . 1895–1896. *Trans. Zool. Soc.,* vol. 15, p. 12 (*Julidochromis ornatus*).
Distribution: Northern region of Lake Tanganyika at Kalungwe, Kashekezi, and Uvira (Mbity Rocks), as well as near Mpulungu at the southern end of Lake Tanganyika.
Habitat: Boulder and rock zones.
First importation into Germany: 1958.
Characters:
fins: D XXII-XXIV/5, A VIII-IX/4–6
scales: mLR 45–50
total length: both sexes up to 85 millimeters.
Sex differences: Scarcely discernible. Males have a small, pointed genital papilla slanted toward the rear.
Aquarium keeping: As in *Julidochromis dickfeldi*.
Spawning: As in *Julidochromis dickfeldi*.

Julidochromis regani Poll, 1942
Striped Slender Cichlid

Explanation of the scientific name: Refers to Ch. T. Regan (1878–1943), the British ichthyologist and museum director of the British Museum of Natural History.

Julidochromis dickfeldi ♂.

Julidochromis dickfeldi, courting.

Julidochromis dickfeldi, courting.

Julidochromis dickfeldi, courting.

Julidochromis dickfeldi, spawning.

Julidochromis dickfeldi, spawning.

Original description: Poll, M. (1942): Cichlidae nouveaux du Lac Tanganyika appartenant aux collections du Musée du Congo. *Rev. Zool. Bot. Afr.*, vol. 36, p. 358.

Distribution: Lake Tanganyika.

Habitat: Boulder and rock zones.

First importation into Germany: 1974.

Characters:

fins: D XXII-XXIII/7–9, A VII-VII/5

scales: mLR 34–35

total length: both sexes up to 130 millimeters.

Sex differences: Scarcely discernible. Males have shorter, more pointed genital papilla slanted toward the rear.

Aquarium keeping: As in *Julidochromis dickfeldi*, although because of the size of the fish, tanks at least 100 centimeters long should be used.

Spawning: As in *Julidochromis dickfeldi*.

Comments: Because of the large area of occurrence various color varieties exist.

Julidochromis transcriptus
Matthes, 1959
Spotted Slender Cichlid

Explanation of the scientific name: Refers to the dark-bordered light spots in the body markings, which appear as if imprinted.

transcribere (L.) = to write over.

Original description: Matthes, H. (1959): Un Cichlide nouveau du lac Tanganyika. *Julidochromis*

transcriptus n. sp. *Rev. Zool. Bot. Afr.*, vol. 60, 1–2, p. 126–130 (*Julidochromis transcriptus*).

Distribution: Northwest shore of Lake Tanganyika between Luhanga and Makabola (Zaire).

Habitat: Boulder and rock zones.

First importation into Germany: 1964.

Characters:

fins: D XXII-XXIII/5–6, A VII-IX/5

scales: mLR 31–36

total length: both sexes up to 70 millimeters.

Sex differences: Scarcely discernible. Males have a small, pointed genital papilla slanted toward the rear.

Aquarium keeping: As in *Julidochromis dickfeldi*.

Spawning: As in *Julidochromis dickfeldi*.

Comments: The markings of the fish in some cases vary greatly depending on the collecting site.

Genus *Lamprologus* Schilthuis, 1890

Explanation of the scientific name: Refers to the bright dots on the body of the type specimen.

lampros (Gk.) = shining, gleaming; *logos* (Gk.) = expression.

Original description: Schilthuis, L. (1890): On a collection of fishes from the Congo; with descriptions of some new species. *Tijdschr. Nederl. Dierk. Verl.*, Ser. 2, III, p. 85.

Distribution: All species come from the Congo (Zaire) River basin.

Number of species: Following the revision of the genus by Colombe and Allgayer (1985), the genus is now represented by five species.

Generic characters: 30 to 53 scales in the middle longitudinal row.

Total length: 90 to 150 millimeters.

Type species: *Lamprologus congoensis* Schilthuis, 1890.

Comments: The former genus *Lamprologus* with its more than fifty species was divided into the following genera by Colombe and Allgayer (1985) on the basis of detailed examination:

Lamprologus Schilthuis, 1890
Lepidiolamprologus Pellegrin, 1904
Neolamprologus Colombe and Allgayer, 1985
Paleolamprologus Colombe and Allgayer, 1985
Variabilichromis Colombe and Allgayer, 1985

Lamprologus congoensis
Schilthuis, 1890

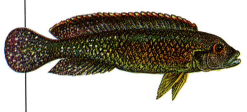

Explanation of the scientific name: Refers to the endemic range of the fish in the Congo (Zaire) River basin.

Original description: Schilthuis, L. (1890): On a collection of fishes from the Congo; with descriptions of some new species. *Tijdschr. Nederl. Dierk. Verl.*, Ser. 2, III, p. 85 (*Lamprologus congoensis*).

Distribution: Area of rapids of the Congo (Zaire) at Matadi (lower Congo), Kinshasa (Stanley Pool), Mosombe (upper Congo), and Stanley Falls.

Julidochromis marlieri.

Julidochromis ornatus.

Julidochromis regani, light form.

Julidochromis regani, dark form.

Julidochromis transcriptus.

Julidochromis transcriptus.

Nanochromis nudiceps.

Nanochromis nudiceps, color variety, described as Nanochromis parilus.

Nanochromis nudiceps, courting.

Nanochromis nudiceps excavating the spawning cavity.

Nanochromis nudiceps, ♀ cleaning the clutch.

Lamprologus congoensis.

First importation into Germany: 1957.

Characters:

fins: D XVIII-XIV/8–10, A VI-VII/5–6

scales: mLR 42–53

total length: male up to 130 millimeters; female up to 100 millimeters.

Sex differences: Males have gleaming dots on the sides of the body. In old age they exhibit a forehead hump.

Aquarium keeping: Fairly large aquaria (from 80 centimeters long) are to be used, the furnishing of which should correspond to the size of the fish and their behavior (numerous rock structures with caves). The hardness of the water is not of prime importance. Important is oxygen-rich water, which one best achieves through the use of a powerful filter pump or, if a centrifugal pump is used, through the use of an injector to return the filtered water to the tank. As a substrate one uses fine-grained to coarse gravel. The tank is planted with *Bolbitis heudelotti* or Java fern. One can keep one male with several females in a species tank. It is also possible to keep this species in a community tank.

Spawning: Females occupy caves, and when they are ready to spawn they lure a male into this cavity. Here the eggs, which are about two millimeters in diameter, are then laid on the roof of the cave. The clutch, which can consist of up to 100 eggs, is cared for by the female. The larvae hatch after 2-1/2 days at a water temperature of about 25° C. Eight days later the fry become free-swimming. The male spawns with several females and guards the entire territory, within which the females possess their own territories.

Lamprologus lethops
Roberts and Stewart, 1976

Explanation of the scientific name: Refers to the appearance of the fish, which look like dead fish.

letum (L.) = death.

Original description: Roberts, T., and Stewart, D. J. (1976): An ecological and systematic survey of the fishes in the rapids of the lower Zaire or Congo River. *Bull. Mus. Comp. Zool., Harvard Univ.* Cambridge, Massachusetts, USA, 147, p. 239–317 (*Lamprologus lethops*).

Distribution: Lower course of the Congo (Zaire) at Balu.

First importation into Germany: Has not yet been imported alive.

Characters:

fins: D XIX/8, A V-VI/6

total length: male up to 100 millimeters; female up to 80 millimeters.

Comments: The fish apparently are blind, a parallel development to the Blind Cave Characin. The body is virtually cylindrical, has comparatively small scales, and exhibits no pigmentation.

Lamprologus mocquardii
Pellegrin, 1903

Explanation of the scientific name: Refers to the proper name Mocquard.

Original description: Pellegrin, J. (1903): *Bull. Mus. Paris*, p. 221 (*Lamprologus mocquardii*).

Distribution: In the region of the Congo (Zaire) and in the upper course of the Ubangi.

First importation into Germany: Apparently has not yet been imported alive.

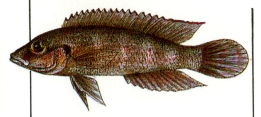

Characters:

fins: D XIX/8, A V-VI/6–7

scales: mLR 22–24

total length: male up to 120 millimeters; female up to 100 millimeters.

Sex differences: Scarcely discernible by external characters.

Aquarium keeping: As in *Lamprologus congoensis*.

Spawning: As in *Lamprologus congoensis*.

Comments: This species is very similar in appearance to *Lamprologus congoensis*, except that the golden dots on the sides of the body are lacking.

Lamprologus werneri
Poll, 1959

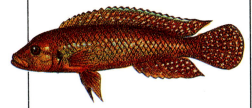

Explanation of the scientific name: Refers to the collector of this species, A. Werner (Munich).

Original description: Poll, M. (1959): Recherches sur la faune ichthyologique de la région du Stanley Pool, *Ann. Mus. Royal, Congo Belge, Sci. Zool.*, vol. 71, p. 75–174 (*Lamprologus werneri*).

Distribution: Stanley Pool and the rapids in the Congo (Zaire) near Kinshasa.

First importation into Germany: 1957.

Characters:

fins: D XIX/9, A VI/6

scales: mLR 28

total length: male up to 120 millimeters; female up to 100 millimeters.

Sex differences: Males have extremely elongated dorsal and

anal fins in the area of the soft fin rays.

Aquarium keeping: As in *Lamprologus congoensis*.

Spawning: As in *Lamprologus congoensis*.

Comments: These fish sometimes have been mistakenly imported under the name *Lamprologus congoensis*.

Genus *Nanochromis* Pellegrin, 1904

Explanation of the scientific name: Refers to the size of the fishes and to the former name for cichlids, "chromides."

nanus (L.) = dwarf; *chroma* (Gk.) = color.

Original description: Pellegrin, J. (1904): Contribution à l'étude anatomique, biolgique et toxonomique des poissons de la famille des Cichlides. *Mem. Soc. Zool. France*, 16, p. 273.

Distribution: Western Africa.

Number of species: So far 14 species are known.

Generic characters: The upper lateral line in part runs on the scales of the base of the dorsal fin.

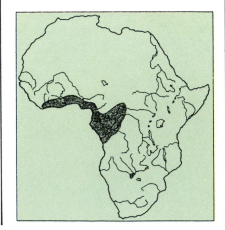

Total length: 70 to 130 millimeters.

Type species: *Nanochromis nudiceps* (*Pseudoplesiops nudiceps* Boulenger, 1899).

Comments: Of the 14 known species, only a few species have been introduced into the aquarium hobby.

Nanochromis caudifasciatus
(Boulenger, 1913)

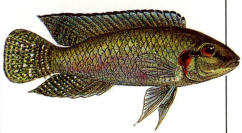

Explanation of the scientific name: Refers to the markings in the tail fin.

cauda (L.) = tail; *fasciatus* (L.) = striped.

Other name in the literature: *Pelmatochromis caudifasciatus* Boulenger, 1913.

Original description: Boulenger, G. A. (1913): *Ann. Mag. Nat. Hist.*, ser. 8, vol. 12, p. 69 (*Pelmatochromis caudifasciatus*).

Distribution: In southern Cameroon, the tributaries of the Nyong River at Pauma, the Nyong River at Akonolinga, the Bumbe River at Assobam, and the Dja River at Bitye and Efayong.

Habitat: The fish were caught near shore under overhanging plants. The water was clear and brownish; the substrate consisted of clayey gravel. The water temperature was 22° C, the total hardness of the water was under 1° dH, and the *p*H was about 6.

First importation into Germany: 1975 by Otto Gartner (Vienna).

Characters:

fins: D XIV-XVI/9–11, A III/7–8

total length: male up to 110 millimeters; female up to 80 millimeters.

Sex differences: Males are more colorful, with dorsal and anal

Nanochromis caudifasciatus.

Nanochromis longirostris.

Nanochromis minor.

Nanochromis spec. aff. minor ♀.

fins tapered to a point. Females have a narrow dark longitudinal stripe extending from the margin of the eye into the tail fin.

Aquarium keeping: It has proved effective to keep the fish in fairly large, at least 70-centimeter-long aquaria, in which the substrate should consist of fine-grained gravel (grain size from 2 to 3 millimeters). The tank should be well planted, and the decoration can consist of resinous bog wood. Since the fish like to take shelter in cavities, one can install a coconut shell or a ceramic flowerpot with a suitable entrance hole in sheltered places in the tank. The water hardness is of no significance for keeping. For spawning, however, it should not be too high. A water hardness of about 10° dH is suitable. The water temperature can be between 22 and 26° C.

Spawning: The tank should be furnished in the same way as for keeping the fish. The water hardness must, however, be maintained at an even lower level. In particular, the carbonate hardness must not be over 2° dH. The fish spawn in a cave. The clutch is cared for by the female, while the male guards the vicinity of the cavity. The female appears outside of the cave with the fry after about ten days. Now the male also takes part in the further care of the brood. The fry are fed with freshly-hatched *Cyclops* or *Artemia*. Frequent water changes are desirable.

Nanochromis dimidiatus
(Pellegrin, 1900)

Explanation of the scientific name: Refers to the differing coloration of the fish above and below the lateral line in fright coloration.

dimidiatus (L.) = to halve.

Other names in the literature: *Pelmatochromis dimidiatus* Pellegrin, 1900

Nannochromis dimidiatus Boulenger, 1915.

Original description: Pellegrin, J. (1900): Poissons nouveaux ou rares du Congo Français. *Bull. Mus. Paris*, p. 99 (*Pelmatochromis dimidiatus*).

Distribution: Ubangi River at Banhi in Zaire.

First importation into Germany: 1952.

Characters:
fins: D XVII/8, A III/6
scales: mLR 25
total length: male up to 80 millimeters; female up to 60 millimeters.

Sex differences: Dorsal and anal fins of males tapered to a point; females have a dark, almost black eyespot in the dorsal fin and a small white spot to the side of the anal opening.

Aquarium keeping: As in *Nanochromis caudifasciatus.*

Spawning: As in *Nanochromis caudifasciatus.*

Comments: This very beautiful species has not been imported in the last fifteen years.

Nanochromis minor
Roberts and Stewart, 1976

Explanation of the scientific name: Refers to the small size of the fish.

minor (L.) = smaller.

Original description: Roberts, T. R., and Stewart, D. J. (1976): An ecological and systematic survey of fishes in the rapids of the lower Zaire or Congo River. *Bull. Mus. Comp. Zool.* Harvard Univ., Cambridge, Massachusetts, USA, 147, p. 289–291 (*Nanochromis minor*).

Distribution: River basin of the lower Congo (Zaire) in Zaire.

Habitat: Narrow tributaries of

the Congo in quiet, rock-rich sites; the fast-flowing bodies of water are very rich in oxygen and had a water temperature of about 25° C (July) in the collecting area. The *p*H was relatively high for this region and was between 7 and 7.5.

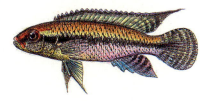

Sex differences: Dorsal and anal fins of males extend to a point. At spawning time females have a yellow throat area and a reddish, swollen belly area; the upper part of the tail fin is also reddish.

Aquarium keeping: As in *Nanochromis caudifasciatus.*

Spawning: As in *Nanochromis caudifasciatus.*

Nanochromis nudiceps
(Boulenger, 1899)

Explanation of the scientific name: Refers to the forehead of the fish.

nudus (L.) = naked, bare;
kephale (Gk.) = head.

Other names in the literature: *Pseudoplesiops nudiceps* Boulenger, 1899

Nannochromis nudiceps Boulenger, 1915.

Original description: Boulenger, G. A. (1899): Matériaux pour la faune du Congo. Poissons nouveaux du Congo. *Ann. Mus. Congo, Zool.* ser. 1, p. 122 (*Pseudoplesips nudiceps*).

Distribution: Upper Congo River basin (Stanley Pool).

Head study of a male *Nanochromis nudiceps*. Below: The pair waltzing about each other. When the female develops her purple belly, you know that she is in breeding condition and will be acceptable to the male.

First importation into Germany: 1952.

Characters:
fins: D XVIII-XIX/8, A III/7
scales: mLR 28–29
total length: male up to 80 millimeters; female up to 50 millimeters.

Sex differences: Males are conspicuously more slender with dorsal and anal fins tapered to a point. Females have a more strongly curved belly area, which is particularly striking in ripe females. The genital papilla is almost always visible.

Aquarium keeping: As in Nanochromis caudifasciatus. The fish are heavy diggers if they are not given a cave from the start.

Spawning: As in Nanochromis caudifasciatus. Females often literally close themselves off from the outside world, in that they pile up sand or gravel from the inside in front of the cave's entrance and seal it off completely.

Comments: Roberts and Stewart (1976) described the new species Nanochromis parilus on the basis of the different coloration of the fins. With many aquarists and in my own spawnings it became apparent that both color forms can appear in the offspring. Since this occurs over several generations, it is questionable whether Nanochromis parilus is truly an independent species.

Nanochromis transvestitus
Stewart and Roberts, 1984

Explanation of the scientific name: Refers to the conspicuously more attractive coloration of the female in comparison to the other Nanochromis species. According to Stewart and Roberts (1984): reversed sexual dichromatism.
vestitus (L.) = clothing, garment; transvestitus = to dress in the garments of the opposite sex.

Original description: Stewart, D. J., and Roberts, T. R. (1984): A new species of Dwarf Cichlid with reversed sexual dichromatism

The female Nanochromis nudiceps tending her eggs which hang from the ceiling of her cave. The eggs are very yellow and hang loosely inside a diaphanous egg case. One end of the egg is white and this is the area in which the embryo will begin to develop. The yellow part of the egg is mostly stored food.

from Lac Mai–Ndombe, Zaire. *Copeia,* nr. 1, p. 82–86 (*Nanochromis transvestitus*).

Distribution: Lac Mai-Ndombe (Lake Leopold II) near Ipeke (Zaire).

Habitat: Shallow, branched-out lake with dark, tea-colored water; depth of visibility not more than 30 centimeters. The *pH* was only 4 at the time of capture.

First importation into Germany: 1985.

Characters:

fins: D XVII/6–7, A III/5, P 12–14

scales: mLR 25–26

total length: male up to 75 millimeters; female up to 55 millimeters.

Sex differences: Males exhibit slightly elongated dorsal and anal fins. The coloration of the entire body is olive-brown with wide dark vertical bars. At spawning time the female is cherry-red colored in the belly region and over a part of the flanks. The soft-rayed parts of the dorsal and anal fins as well as the tail fin exhibit white bands on their dark ground color.

Aquarium keeping: As in *Nanochromis caudifasciatus.*

Spawning: As in *Nanochromis caudifasciatus.*

Genus *Neolamprologus* Colombe and Allgayer, 1985

Explanation of the scientific name: Refers to the separation from the genus *Lamprologus.* *Neos* (Gk.) = new.

Original description: Colombe, J., and Allgayer, R. (1985): Description de *Variabilichromis, Neolamprologus* et *Paleolamprologus* genres nouveaux du Lac Tanganyika, avec description des genres *Lamprologus* Schilthuis, 1891 et *Lepidiolamprologus* Pellegrin, 1904 (Pisces: Teleostei: Cichlidae). *Rev. Franc. Cich.,* nr. 5, p. 9–16 and 21–28.

Distribution: Lake Tanganyika (Africa).

Number of species: According to the revision by Colombe and Allgayer (1985), 39 species belong to the genus.

Generic characters: No more than 60 scales in the central longitudinal row; all species come from Lake Tanganyika.

Total length: 40 to 200 millimeters.

Type species: *Neolamprologus tetracanthus* (*Lamprologus tetracanthus* Boulenger, 1899).

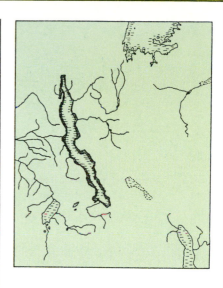

This female *Nanochromis transvestitus* is in her spawning garb.

The ordinary colored *Nanochromis transvestitus* doesn't look like anything special. ▲

The male displays before the female. ▼

The female, when in spawning condition, shows a cherry-red tinged belly. ▲

The female displays her own swollen belly to the male. ▼

Pretty soon the two fish are so similarly colored that you don't know which is the male and which is the female (thus the scientific name *transvestitus*). ▼

It is only the red belly of the female (shown in the background) that allows you to differentiate between the two specimens. ▼

The female continues to act like a male, displaying with all her vigor. ▼

When an intruder approaches, the female might well attack it as is shown here. ▼

Neolamprologus brevis
(Boulenger, 1899)

Explanation of the scientific name: Refers to the small size of this species.

brevis (L.) = short, small.

Other name in the literature: *Lamprologus brevis* Boulenger, 1899.

Original description: Boulenger, G. A. (1899): Matériaux pour la faune du Congo. *Poissons nouveaux du Congo. Ann. Mus. Congo, Zool.*, 1, p. 115 (*Lamprologus brevis*).

Distribution: Lake Tanganyika.

Habitat: The fish exclusively stay near so-called snail graveyards; that is, in places where a fairly large number of empty snail shells of the genus *Neothauma* can be found. They lie in depths of between 6 and 50 meters. The empty snail shells serve as shelters and spawning sites.

First importation into Germany: 1979.

Characters:

fins: D XVI-XIX/7–9, A VII-IX/5–7

scales: mLR 30–35

total length: male up to 55 millimeters; female up to 40 millimeters.

Sex differences: Males are more colorful, with dorsal fin tapering to a point. Females are mostly recognizable by the fuller and lighter belly area.

Aquarium keeping: For these small, extremely shy fish one needs only a relatively small aquarium. Nevertheless, it should not be smaller than 50 by 30 by 30 centimeters. One furnishes the tank with fine-grained gravel or sand and plants it with *Anubias* or *Bolbitis*. In the foreground one places several empty snail shells. Medium-hard water is best suited for keeping these fish. It can have a temperature of between 23 and 28° C. In a tank furnished in this manner one can keep one or two pairs. The fish usually disappear into the empty snail shells after they are placed in the tank. If possible, the fish should be fed with live food, of which *Cyclops* is the most suitable. However, Grindal worms, whiteworms, *Daphnia*, and *Tubifex* are also eaten.

Spawning: For spawning one does not have to alter the keeping tank. The fish turn the snail shell in the desired direction with their mouths and bury it in such a way that finally only the opening is visible. When alarmed, the fish then swim into the shell in a flash. Ready-to-spawn females are easily recognized by their swollen, light belly area. Just before spawning they court constantly with spread fins in front of the male, which then almost exclusively stays in the immediate vicinity of the snail shell and looks into the opening with spread fins again and again. Once the female lays the eggs on the interior wall of the snail shell, the male draws near and positions itself with fins flattened against its body at the opening. Here the male releases the sperm, which apparently reaches the clutch through the suction produced by the female as it swims in and out of the opening. The brood is subsequently cared for exclusively by the female. The fry, at first only about two millimeters long, return again and again to the snail shell during the first few days. They are not led by the parents.

Comments: Because of the large area of occurrence, there are various color varieties of this species.

Neolamprologus buescheri
Staeck, 1893

Explanation of the scientific name: Refers to H. H. Buscher, the collector of this species.

Other name in the literature: *Lamprologus buescheri* Staeck, 1983.

Original description: Staeck, W. (1983): *Lamprologus buescheri* n. sp. from Zambian part of Lake Tanganyika (Pisces: Cichlidae). *Senckenbergiana biol.* 63, p. 325–328 (*Lamprologus buescheri*).

Distribution: Cape Kachese, Lake Tanganyika (Zambia).

Habitat: Boulder and rock zones at a depth of 16 to 18 meters.

First importation into Germany: 1982 by H. H. Buscher.

Characters:

fins: D XVIII/9, A VI/7

scales: mLR 36

total length: male up to 80 millimeters; female up to 70 millimeters.

Sex differences: Not distinguishable by external characters.

Aquarium keeping: Tanks with a length of at least 80 centimeters should be used. The substrate should consist of fine-grained gravel. Rock structures with cavities must not be lacking in any case. The tank can be planted with *Anubias* species. The water hardness does not play a role; however, the pH should not be under the neutral value of 7. The water temperature can be between 23 and 28° C.

Spawning: The fish spawn in cavities. The clutch is principally cared for by the female. The parents do not continue to tend the fry; however, the territory in which the fry stay is guarded intensively by the parents.

Neolamprologus brevis ♂.

Neolamprologus brevis, ♀ displaying in front of the ♂.

Neolamprologus brevis, ♂ fertilizes the clutch located in the snail shell.

Hybrid between *Julidochromis marlieri* and *Neolamprologus elongatus*.

Neolamprologus elongatus Daffodil.

Neolamprologus elongatus.

Neolamprologus buescheri.

Neolamprologus elongatus
(Trewavas and Poll, 1952)
Princess of Burundi Fairy Cichlid

Explanation of the scientific name: Refers to the form of the fins.

elongatus (L.) = elongated.

Other names in the literature:
Lamprologus savoryi elongatus Trewavas and Poll, 1952

Lamprologus brichardi Poll, 1974.

Original description: Trewavas, E., and Poll, M. (1952): Three new species and two new subspecies of the genus *Lamprologus Bull. Inst. Royal Sci. Nat. Belgique, Bruxelles*, t. 28, nr. 50, p. 5–6 (*Lamprologus savoryi elongatus*).

Distribution: Northern half of Lake Tanganyika.

Habitat: Boulder and rock zones at a depth of 5 to 10 meters.

First importation into Germany: 1970.

Characters:
fins: D XIX-XX/8–9, A V-VII/5–7
scales: mLR 32–36
total length: Both sexes up to 90 millimeters.

Sex differences: No external characters.

Aquarium keeping: As in *Neolamprologus buescheri*.

Spawning: As in *Neolamprologus buescheri*.

Comments: This attractive species has been known up till now by the name of *Lamprologus brichardi*. In the revision of the genus *Lamprologus* by Colombe and Allgayer (1985) it turned out that *Lamprologus savoryi elongatus*, initially classified as a subspecies, is an independent species, and according to the rules of nomenclature the specific name *elongatus* is valid.

Neolamprologus elongatus was previously well known under the name *Lamprologus brichardi*. The pair "making love" have already deposited a few eggs on the stone below.

A pair of *Neolamprologus elongatus* preparing to spawn. The male is the fish with the longer fins.

On account of the large area of occurrence one finds various color varieties in this species. Recently the variety designated by the name "daffodil" has become very popular because of its more beautiful coloration.

Neolamprologus hecqui
(Boulenger, 1899)

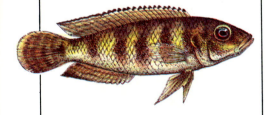

Explanation of the scientific name: Refers to Capt. Hecq, collector of the type.
Other names in the literature:
Lamprologus hecqui Boulenger, 1899
Julidochromis boulengeri Steindachner, 1909.
Original description: Boulenger, G. A. (1899): Matériaux pour la faune du Congo. Poissons nouveaux du Congo. *Ann. Mus. Congo, Zool.,* 1, p. 115 (*Lamprologus hecqui*).
Distribution: Lake Tanganyika (Mtondwe Bay, Niamkolo Bay, Kalembwe).
Habitat: Sandy and muddy substrate in the immediate vicinity of empty snail shells of the genus *Neothauma.*
First importation into Germany: 1985 by W. Dieckhoff (Herten).
Characters:
  fins: D XVIII–XIX/8–9, A VI–VII/7–8
  total length: male up to 80 millimeters; female up to 40 millimeters.
Aquarium keeping: As in *Neolamprologus brevis.*
Spawning: As in *Neolamprologus brevis.*

Neolamprologus kungweensis
(Poll, 1956)

Explanation of the scientific name: Refers to the collecting area, Kungwe Bay (Tanzania).
Other names in the literature:

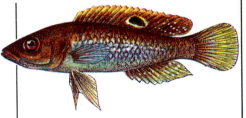

Lamprologus ocellatus Poll, 1952
Lamprologus kungweensis Poll, 1956.
Original description: Poll, M. (1956): Poissons Cichlidae. *Résultats scientifiques exploration hydrobiologique du Lac Tanganyika 1946–1947.* vol. 3, ser. 5 B, p. 570–571 (*Lamprologus kungweensis*).
Distribution: Lake Tanganyika, Kungwe Bay and at Kigoma, on the northeastern shore.
Habitat: Sandy to muddy substrate at a depth of about 20 meters with accumulations of empty snail shells of the genus *Neothauma.*
First importation into Germany: 1985 by W. Dieckhoff (Herten).

Neolamprologus cylindricus.

Nanochromis dimidiatus ♀.

Nanochromis dimidiatus ♂.

Characters:
fins: D XIV-XV/8–9, A VI/6
scales: mLR 30–33
total length: male up to 80 millimeters; female up to 40 millimeters.

Sex differences: The dorsal and anal fins of the male are tapered to a point. The much smaller, yellow-brown female has a dark spot in the dorsal fin.

Aquarium keeping: As in *Neolamprologus brevis*.

Spawning: As in *Neolamprologus brevis*.

Neolamprologus leleupi
(Poll, 1956)

Explanation of the scientific name: Refers to the proper name Leleup.

Other names in the literature:
Lamprologus leleupi Poll, 1956
Lamprologus leleupi melas Matthes, 1959
Lamprologus leleupi longior Staeck, 1980.

Original description: Poll, M. (1956): Poissons Cichlidae. *Résultats scientifiques exploration hydrobiologique du Lac Tanganyika 1946–1947.* vol. 3, ser. 5 B, p. 1–619 (*Lamprologus leleupi*).

Distribution: Southern half of Lake Tanganyika.

Habitat: Boulder and rock zones at depths of between 3 and 30 meters.

First importation into Germany: 1958.

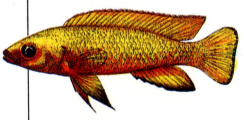

Characters:
fins: D XIX-XX/8, A V-VI/6
scales: mLR 31–34
total length: male up to 100 millimeters; female up to 80 millimeters.

Sex differences: Other than the size, no distinct external characters.

Aquarium keeping: As in *Neolamprologus buescheri*.

Spawning: As in *Neolamprologus buescheri*.

Comments: Three subspecies of this species are described, which in my opinion are not sufficiently distinct from one another to be considered as such. In all likelihood it is a question here of fish of one and the same species that occur in different color varieties because of the large area of occurrence, but which are designated as subspecies by Matthes (1959) and Staeck (1980).

Neolamprologus leloupi
(Poll, 1948)

Explanation of the scientific name: Refers to E. Leloup, the director of the Lake Tanganyika Hydrobiological Mission.

Other name in the literature:
Lamprologus leloupi Poll, 1948.

Original description: Poll, M. (1948): *Bull. Mus. Royal Hist. Nat.* Belg., XXIV, 26, p. 27 (*Lamprologus leloupi*).

Distribution: Lake Tanganyika.

Habitat: Boulder and rock zones.

First importation into Germany: Has not been imported alive so far.

Characters:
fins: D XVII/7, A VI/6
scales: mLR 31
total length: type specimen 61 millimeters.

Neolamprologus meeli (Poll, 1948)

Explanation of the scientific name: Refers to M. L. van Meel, member of the Lake Tanganyika Hydrobiological Mission.

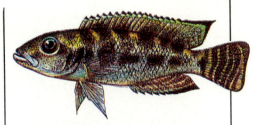

Other name in the literature:
Lamprologus meeli Poll, 1948.

Original description: Poll, M. (1948): *Bull. Mus. Royal Hist. Nat. Belg.,* XXIV, 26, p. 29 (*Lamprologus meeli*).

Distribution: Katibili Bay in Lake Tanganyika (delta of the Kokivi River).

Habitat: Sandy substrate with accumulations of empty snail shells of the genus *Neothauma*.

First importation into Germany: 1980.

Characters:
fins: D XVIII/9–10, A VII/7–8
scales: mLR 42–46
total length: male up to 70 millimeters; female up to 40 millimeters.

Sex differences: Males are more colorful with more intense mottling on the body.

Aquarium keeping: As in *Neolamprologus brevis*.

Spawning: As in *Neolamprologus brevis*.

Neolamprologus modestus
(Boulenger, 1898)

Explanation of the scientific name: Refers to the fish's appearance.
modestus (L.) = modest.

Other name in the literature:
Lamprologus modestus Boulenger, 1898.

Neolamprologus leleupi.

Clutch of *Neolamprologus leleupi.*

Moment of hatching of the larvae.

The larvae are gathered together in one spot.

153

Original description: Boulenger, G. A. (1898): Report on the collection of fishes made by Mr. J. E. S. Moore in Lake Tanganyika . . . 1895–1896. *Trans. Zool. Soc., London*, 15, p. 8 (*Lamprologus modestus*).

Distribution: Lake Tanganyika (Mbity Rocks, Kinyamkolo, Nkamba Bay).

Habitat: Higher boulder and rock zones.

First importation into Germany: 1980.

Characters:
fins: D XIX-XX/8–10, A V-VI/6–7
scales: mLR 34–37
total length: male up to 120 millimeters; female up to 100 millimeters.

Sex differences: Only slight external differences. In older males the dorsal and anal fins are more pointed.

Aquarium keeping: As in *Neolamprologus buescheri*.

Spawning: As in *Neolamprologus buescheri*.

Neolamprologus multifasciatus
(Boulenger, 1906)

Explanation of the scientific name: Refers to the numerous vertical bars on the sides of the body of the fish.
multus (L.) = many, numerous; *fasciatus* (L.) = striped.

Other name in the literature: *Lamprologus multifasciatus* Boulenger, 1906.

Original description: Boulenger, G. A. (1906): Fourth contribution to the ichthyology of Lake Tanganyika. Report on the collection of fishes made by Dr. W. A. Cunnington during third Tanganyika expedition, 1904–1905. *Trans. Zool. Soc., London*,

vol. 17, p. 558 (*Lamprologus multifasciatus*).

Distribution: Niamkolo Bay, Sumbu Bay, and Luvu Bay in Lake Tanganyika.

Habitat: Sites up to 10 meters deep with accumulations of empty snail shells of the genus *Neothauma*.

First importation into Germany: 1984.

Characters:
fins: D XVII-XVIII/8–9, A VI-VII/6
scales: mLR 29–31
total length: male up to 50 millimeters; female up to 40 millimeters.

Sex differences: The male exhibits more intensely developed markings and possesses a reddish edging of the dorsal and anal fins.

Aquarium keeping: As in *Neolamprologus brevis*.

Spawning: As in *Neolamprologus brevis*.

Neolamprologus ocellatus
(Steindachner, 1909)

Explanation of the scientific name: Refers to the dark spot on the edge of the gill cover.
ocellatus (L.) = having an eyespot.

Other names in the literature: *Julidochromis ocellatus* Steindachner, 1909
Lamprologus lestradei Poll, 1943
Lamprologus ocellatus Poll, 1946.

Original description: Steindachner, F. (1909): *Anz. Akad. Wiss., Wien*, 46, p. 402 (*Julidochromis ocellatus*).

Distribution: Lake Tanganyika (Tembwe Bay, Kigoma).

Habitat: On sandy or slightly muddy substrates with

accumulations of empty snail shells of the genus *Neothauma* at depths of 5 to 15 meters.

First importation into Germany: 1979.

Characters:
fins: D XVI-XVIII/6–8, A VII-VIII/6–7
scales: mLR 25–30
total length: male up to 55 millimeters; female up to 35 millimeters.

Sex differences: Males have larger fins and are somewhat more colorful.

Aquarium keeping: As in *Neolamprologus brevis*.

Spawning: As in *Neolamprologus brevis*.

Neolamprologus ornatipinnis
(Poll, 1949)

Explanation of the scientific name: Refers to the markings in the fins.
ornatus (L.) = adorned, beautiful; *pinna* (L.) = fin.

Other name in the literature: *Lamprologus ornatipinnis* Poll, 1949.

Original description: Poll, M. (1949): *Bull. Inst. Roy. Sc. Nat. Belg.*, XXV, 33, p. 50 (*Lamprologus ornatipinnis*).

Distribution: Lake Tanganyika (Burton Bay).

Habitat: On sandy to muddy substrates with accumulations of empty snail shells of the genus *Neothauma* at depths of up to 120 meters.

First importation into Germany: 1984.

Characters:
fins: D XV-XVIII/7–9, A V-VIII/6–8
scales: mLR 32–36
total length: male up to 75 millimeters; female up to 40 millimeters.

Neolamprologus ocellatus male selects a shell. ▲

He immediately digs around the shell to anchor it more securely. ▲

He digs fairly deeply. ▲

His digging attracts a female, and the male enters the shell. ▲

Then the female enters the shell. ▲

The female leaves the shell and waits. ▼

The male waits while the female deposits her eggs. ▲

The male enters the shell, fertilizes the eggs, and signals the female to lay more eggs. ▼

The female *Neolamprologus leleupi* tears open the egg case and releases her hatching young, leaving the egg case still attached to the rock. Lower, left: The female tends the eggs while they are developing. Lower, right: The female stows the young in a safe storage area.

Neolamprologus meeli.

Neolamprologus multifasciatus.

Neolamprologus ocellatus.

Sex differences: Males are a uniform yellow-gray with dorsal and anal fins that taper to a point. Females have dark markings on the dorsal fin. In breeding coloration the female is provided with a silver-green spot on the front part of the belly area and the rear part is a dark-brown color. The ventral fins are sooty black.

Aquarium keeping: As in *Neolamprologus brevis.*

Spawning: As in *Neolamprologus brevis.*

Neolamprologus savoryi
(Poll, 1949)

Explanation of the scientific name: Refers to the proper name Savory.

Other name in the literature: *Lamprologus savoryi* Poll, 1949.

Original description: Poll, M. (1949): *Bull. Inst. Roy. Sc. Nat. Belg.,* XXV, 33, p. 52 (*Lamprologus savoryi*).

Distribution: Lake Tanganyika.

Habitat: Boulder and rock zones close to shore in depths of up to 4 meters.

First importation into Germany: 1958.

Characters:
fins: D XVIII-XIX/7–9, A VI-VII/6–7
scales: mLR 33–35, Ltr 26
total length: male up to 85 millimeters; female up to 70 millimeters.

Sex differences: Not distinguishable by external characters.

Aquarium keeping: As in *Neolamprologus buescheri.*

Spawning: As in *Neolamprologus buescheri.*

Neolamprologus wauthioni
(Poll, 1949)

Explanation of the scientific name: Refers to M. R. Wauthion.

Other name in the literature: *Lamprologus wauthioni* Poll, 1949.

Original description: Poll, M. (1949): *Bull. Inst. Roy. Sc. Nat. Belg.,* XXV, 33, p. 47 (*Lamprologus wauthioni*).

Distribution: Katibili Bay in Lake Tanganyika.

Habitat: About one kilometer from shore at a depth of 35 meters on muddy substrates with large accumulations of empty snail shells of the genus *Neothauma.*

First importation into Germany: 1984.

Characters:
fins: D XVII-XVIII/7–9, A VI/6
scales: mLR 30–31
total length: male up to 55 millimeters; female up to 35 millimeters.

Sex differences: Males are yellowish-brown with a light belly and large dark spots on the back and the upper parts of the sides of the body.

Aquarium keeping: As in *Neolamprologus brevis.*

Spawning: As in *Neolamprologus brevis.*

Even when a solitary *Neolamprologus* is in the tank, it will use the inside of a snail shell as a refuge. Many *Neolamprologus* sleep inside their snail shells. For the collector, *Neolamprologus* are easy to catch since you simply lift them into your net, shell and all.

Neolamprologus fasciatus.

Neolamprologus spec.

Neolamprologus staecki.

Neolamprologus spec.

This remarkable series of photos shows two species of *Neolamprologus*, namely *Neolamprologus brevis* and *N. "margarae,"* a still undescribed species. If you look closely at the two photos, upper left, you'll notice an eye-sized golden spot slightly above the rear of the eye. This is the new species.

The spawning sequence is that the female enters the empty shell, lays her eggs, and the male fertilizes them thereafter. As you can see from some of the closeups, these fish have "fangs" which are curved and very sharp. Exactly what function these "canine" teeth serve is still uncertain.

**Genus *Pelvicachromis*
Thys and Loiselle, 1971**

Explanation of the scientific name: Refers to the shape of the female's belly and to the former name for cichlids, "chromides."
Pelvis (L.) = basin, bowl; *chroma* (Gk.) = color.

Original description: Thys van den Audenaerde, D. F. E., and Loiselle, P. V. (1971): Description of two new small African Cichlids. *Rev. Zool. Bot. Afr.,* 83, p. 193–206.

Distribution: Western Africa, from Guinea to Angola.

Number of species: So far eight species are known according to the revision of Thys van den Audenaerde (1964).

Generic characters: The ventral fins of the male and female are differently formed. Females possess rounded-off or sloped ends of the ventral fins, whereas those of males taper to a point.

Total length: Males 70 to 120 millimeters; females 50 to 80 millimeters.

Type species: *Pelvicachromis pulcher* (*Pelmatochromis pulcher* Boulenger, 1901).

Comments: For decades the fishes of this genus went by the name *Pelmatochromis*.

Pelvicachromis humilis
(Boulenger, 1916)

Explanation of the scientific name: Refers to the body

coloration of the fish.
humilis (L.) = simple, lowly.

Other name in the literature: *Pelmatochromis humilis* Boulenger, 1916.

Original description: Boulenger, G. A. (1916): Catalogue of the fresh-water fishes of Africa in the British Museum. *British Mus., London,* vol. 4, p. 333-334 (*Pelmatochromis humilis*).

Distribution: Sierra Leone, Guinea, Liberia.

Habitat: Slow-flowing bodies of water with very soft, light brownish water (total hardness under 1° dH), with a *p*H of about 6 and a water temperature of about 25° C. The substrate consists of volcanic sand with dark gravel and groups of rocks. The fish chiefly stay in sheltered places under overhanging riparian plants and roots.

First importation into Germany: 1978 by H. Linke (West Berlin).

Characters:
fins: D XVII/11, A III/7–8
scales: mLR 20–21
total length: male up to 130 millimeters; female up to 80 millimeters.

Aquarium keeping: The fish should be offered an as-large-as-possible aquarium (not less than 80 centimeters long). The background and sides of the tank should be well planted with *Anubias* species or *Bolbitis heudelotti.* In the foreground one places a cave constructed of rocks or in the form of a coconut shell. As a substrate one uses gravel with a grain size of up to 5 millimeters. The water temperature should be between 23 and 28° C.

Spawning: Soft to medium-

hard water should be used. The fish spawn in a cave. After that the clutch is cared for by the female alone. Only after the female leads the fry out of the cave does the male also take part in brood care.

Pelvicachromis pulcher
(Boulenger, 1901)
Purple Splendid Cichlid

Explanation of the scientific name: Refers to the coloration of the fish.
pulcher (L.) = beautiful, lovely.

Other names in the literature:
Pelmatochromis pulcher Boulenger, 1901
Pelmatochromis aureocephalus Meinken, 1960
Pelmatochromis camerunensis?
Pelmatochromis kribensis Boulenger, 1911
Pelmatochromis subocellatus kribensis?

Original description: Boulenger, G. A. (1901): On the fishes collected by Dr. W. J. Ansorge in the Niger delta. *Proc. Zool. Soc.,* London, vol. 1, p. 9 (*Pelmatochromis pulcher*).

Distribution: The mouth of the Ethiop River, Niger delta; according to Linke, 120 kilometers west of the Niger delta at Bemin City (Nigeria).

Habitat: Small water courses with clear water, in stands of aquatic plants and their immediate vicinity, in places with little current and at depths of up to 60 centimeters. Water hardness less than 1° dH at a *p*H of about 5. Water temperature at the time of collection about 27° C.

First importation into Germany: 1913 by Christian Brunning (Hamburg).

Characters:
fins: D XVI/9–10, A III/7–8
scales: mLR 18–20
total length: male up to 110 millimeters; female up to 75 millimeters.

Sex differences: Males are more slender with dorsal and anal fins tapered to a point. Females have a compact build and a strongly curved belly area that is particularly noticeable at the onset of spawning, when it is a cherry-red color.

Aquarium keeping: As in *Pelvicachromis humilis.*

Spawning: As in *Pelvicachromis humilis.*

Comments: This species includes a large number of color varieties, although according to the revision by Thys van den Audenaerde (1964) two forms are distinguished, the A and the B form. As a rule, the A form possesses either a weak longitudinal stripe on the middle of the body or none at all, whereas the B form always exhibits a dark longitudinal stripe.

Pelvicachromis roloffi
(Thys van den Audenaerde, 1968)
Roloff's Splendid Cichlid

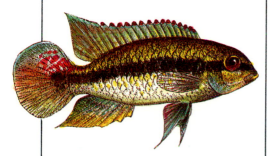

Explanation of the scientific name: Refers to the discoverer of this species, E. Roloff.

Other name in the literature:
Pelmatochromis roloffi Thys van den Audenaerde, 1968.

Original description: Thys van den Audenaerde, D. F. E. (1968): Description of a new *Pelmatochromis* (Pisces, Cichlidae) from Sierra Leone. *Rev. Zool. Bot. Afr.,* vol. 77 (3–4), p. 335–345 (*Pelmatochromis roloffi*).

Distribution: Lake Kwako (Sierra Leone), eastern Guinea, and western Liberia.

First importation into Germany: 1968 by Erhard Roloff (Karlsruhe).

Characters:
fins: D XVI-XVII/9, A III/7, P 12–13
scales: mLR 27–28
total length: male up to 90 millimeters; female up to 60 millimeters.

Sex differences: Dorsal and anal fins of male taper to a point. At spawning time females have an intense violet-colored belly area.

Aquarium keeping: As in *Pelvicachromis humilis.*

Spawning: As in *Pelvicachromis humilis.*

Pelvicachromis subocellatus
(Günther, 1871)
Eyespot Splendid Cichlid

Explanation of the scientific name: Refers to the eyespot in the soft-rayed part of the female's dorsal fin.
sub (L.) = below; *ocellatus* (L.) = having an eyespot.

Other names in the literature:
Hemichromis subocellatus Günther, 1871
Pelmatochromis subocellatus Boulenger, 1898
Pelmatochromis klugei II?

Original description: Günther, A. (1871): Report on several collections of fishes recently obtained for the British Museum. *Proc. Zool. Soc., London,* p. 667 (*Hemichromis subocellatus*).

Distribution: From Libreville as far as Moanda in the lower Congo (Zaire) River basin (Gabon).

First importation into Germany: 1907 by W. Schroot (Hamburg).

Characters:
fins: D XIV-XVI/8–10, A III/6–8
total length: male up to 90 millimeters; female up to 65 millimeters.

Sex differences: The dorsal and anal fins of the male are tapered to a point. The female has a compact build; at spawning time

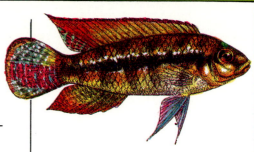

it has a thick, light red belly area, in part with a lighter ring, which is edged in almost black areas during courtship and shortly before spawning.

Aquarium keeping: As in *Pelvicachromis humilis.*

Spawning: As in *Pelvicachromis humilis.*

Comments: Two color varieties of this species are known, and within the form formerly designated as *Pelmatochromis klugei II* even more color nuances occur.

Pelvicachromis taeniatus
(Boulenger, 1901)
Emerald Splendid Cichlid

Explanation of the scientific name: Refers to the longitudinal stripe on the middle of the body.
taenia (L.) = stripe.

Other names in the literature:
Pelmatochromis taeniatus Boulenger, 1901
Pelmatochromis klugei Meinken, 1960
Pelmatochromis kribensis klugei Meinken, 1965.

Original description: Boulenger, G. A. (1901): On the fishes collected by Dr. W. J. Ansorge in the Niger delta. *Proc. Zool. Soc., London,* vol. 1, p. 10 (*Pelmatochromis taeniatus*).

Distribution: According to Boulenger (1901), the Niger delta; according to Linke (1981), other color varieties occur in Cameroon.

Habitat: Small water courses, in part with aquatic plant growth. The fish chiefly stay among branches in the water, exposed roots, or aquatic plants. Almost everywhere the water has a total hardness of not more than 5° dH at *p*H values of between 5 and 7.5 The water

Pelvicachromis taeniatus spawning (above). The characteristic rose-colored belly of the female is easily observed. The female (below) guards the developing eggs.

A pair of *Pelvicachromis taeniatus* with their approximately one month old fry. The family stays together very successfully if they are maintained in their own aquarium.

Pelvicachromis taeniatus ♂, Nigeria form.

Pelvicachromis taeniatus ♀.

Pelvicachromis subocellatus and *Pelvicachromis taeniatus*, fighting males.

Pelvicachromis taeniatus, Kienke form.

Pelvicachromis taeniatus, Lobe form.

Pelvicachromis taeniatus, Moliwe form.

Pelvicachromis spec. aff. *subocellatus* ♀.

Pelvicachromis subocellatus ♂.

Pelvicachromis subocellatus ♀.

Pelvicachromis taeniatus, head study.

temperature was about 26° C at the time of collection.

First importation into Germany: 1911 by Christian Brünning (Hamburg).

Characters:
fins: D XVII-XVIII/7–8, A III/7
scales: mLR 28–29
total length: male up to 90 millimeters; female up to 60 millimeters.

Sex differences: Males are more slender and have dorsal and anal fins tapered to a point. Females have the typical swollen belly area, particularly at spawning time.

Aquarium keeping: As in *Pelvicachromis humilis*.

Spawning: As in *Pelvicachromis humilis*.

Comments: With this species, depending on the locality, one finds different color varieties, of which, however, only a few survived in aquaria after the first importation.

Pelvicachromis humilis, Kenema color form.

Pelvicachromis humilis, Kasewe color form.

Pelvicachromis subocellatus, color variety.

Pelvicachromis subocellatus, color variety.

Pelvicachromis subocellatus ♀, yellow color form.

Pelvicachromis pulcher, color variety.

Pelvicachromis pulcher, color variety.

Pelvicachromis pulcher, color variety.

Pelvicachromis roloffi.

Pelvicachromis pulcher, color variety.

Genus *Steatocranus*
Boulenger, 1899

Explanation of the scientific name: Refers to the head of the fishes.

steatos (Gk.) = fat; *cranium* (L.) = skull.

Original description: Boulenger, G. A. (1899): Matériaux pour la faune du Congo. Poissons nouveaux du Congo. *Ann. Mus. Congo, Zool.,* vol. 1, p. 52.

Distribution: Zaire and Ghana.

Number of species: So far eight species are known.

Generic characters: These are elongated fishes that have adapted to life in the rapids and that chiefly stay on the bottom. As a result of the vestigial swim bladder, the principal mode of locomotion does not consist of swimming, but rather more of a sort of hopping, in the course of which the fishes prop themselves on the bottom with their ventral fins and hold fast to the substrate. Older males exhibit a fatty pad on the forehead that lends the fishes a blunt-headed appearance. The dorsal fin has 19 or 20 spines and the anal fin 3 spines.

Total length: 100 to 150 millimeters.

Type species: *Steatocranus gibbiceps* Boulenger, 1899.

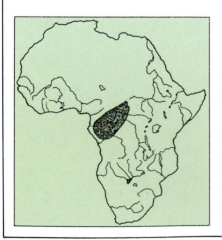

Steatocranus casuarius
Poll, 1939
Humphead Cichlid

Explanation of the scientific name: Refers to the forehead hump of males.

cassida (L.) = helmet.

Other name in the literature: *Steatocranus elongatus* The Aquarium, 1956.

Original description: Poll, M. (1939): Les poissons du Stanley-Pool. *Ann. Mus. Congo Belge, Zool.,* ser. 1, vol. 4 (1), p. 1–60 (*Steatocranus casuarius*).

Distribution: Lower course of the Congo (Zaire) and its tributaries.

Habitat: Rapids and fast-flowing bodies of water. The fish live in still-water zones on the bottom where they seek out places to stay among boulder fragments. Water temperature between 25° and 29° C.

First importation into Germany: 1956 by the firm of Werner (Munich).

Characters:
fins: D XIX-XX/8; A III/6
total length: male up to 120 millimeters; female up to 100 millimeters.

Sex differences: Males are more blunt-headed in appearance. The forehead hump of the male is conspicuously larger than in the female, as are the ventral fins.

Aquarium keeping: Corresponding to the size of the fish, one should offer them a large tank with a length of over 100 centimeters. Because the fish, particularly when they do not find any suitable caves, dig a great deal, it is very important to build rock structures with caves for them. The rock structures must stand directly on the bottom of the aquarium, so that they do not collapse during digging. Gently-flowing, oxygen-rich water contributes to their well-being. This is best achieved through the use of a powerful filter pump and an injection nozzle to return the filtered water to the tank. If sufficient rock structures are present, it is also possible to introduce aquatic plants. If possible, they should be planted in groups. One can keep other fishes with them in the tank; however, they should be species that favor the upper water levels.

Spawning: Spawning the fish presents relatively few problems. They spawn in a cave and usually attach the approximately three-millimeter-long eggs to the roof of the cave. After spawning, the female alone cares for the clutch. The larvae hatch after seven days. The larvae are then tended by the female for an additional ten days in the cave. Once the fry leave the cavity after this time, the male also takes part in brood care. Frequent water changes are now particularly beneficial.

Steatocranus tinanti (Poll, 1939)

Explanation of the scientific name: Refers to the proper name Tinant.

Other names in the literature: *Gobiochromis tinanti* Poll, 1939 *Leptotilapia tinanti* Trewavas and Irvine, 1947.

Original description: Poll, M. (1939): Les poissons du Stanley-Pool. *Ann. Mus. Congo Belge, Zool.,* ser. 1, vol. 4 (1), p. 1–60 (*Gobiochromis tinanti*).

Distribution: Lower course of the Congo (Zaire) rapids at Kinshasa (Zaire).

Habitat: Fast-flowing bodies of water with sandy substrates, in which are interspersed rocks of all kinds.

First importation into Germany: 1958.

Characters:
fins: D XX-XXI/8, A III/6
total length: male up to 120 millimeters; female up to 100 millimeters.

Sex differences: Dorsal fin of male tapered to a point; head blunter.

Aquarium keeping: As _Steatocranus casuarius_.

Spawning: As _Steatocranus casuarius_.

**Genus _Taeniacara_
Myers, 1935**

A closeup of the face of _Steatocranus casuarius_ clearly showing the fatty lump on the forehead and the enlarged lips. The bump on the forehead may protect the eyes of the fish, which prefers fast-moving waters with a bottom strewn with stones.

The face of *Steatocranus tinanti* showing teeth specialized in scraping algae from rocks and wood. Enlarged lips are usually developed. According to Dr. Herbert R. Axelrod, when young *Steatocranus* are maintained in an all-glass aquarium and fed a normal aquarium diet, their lips do not develop as much as the lips found on wild specimens.

Explanation of the scientific name: Refers to the conspicuous dark longitudinal stripe on the sides of the body.

taenia (L.) = stripe, band; *acara* = native name for cichlids.

Original description: Myers, G. S. (1935): Four new freshwater fishes from Brasil, Venezuela and Paraguay. *Proc. Biol. Soc., Washington,* vol. 48, p. 11.

Distribution: Region of the upper Rio Negro.

Number of species: So far only one species is known.

Generic characters: The lateral line is lacking or is only suggested.

Total length: Up to 70 millimeters.

Type species: *Taeniacara candidi* Myers, 1935.

Taeniacara candidi Myers, 1935

Explanation of the scientific name: Refers to E. Candidus, who gave the fish to G. S. Myers for description.

Other names in the literature:
Apistogramma spec. 2 1914 Arnold, 1914
Apistogramma weisei Ahl, 1936.

Original description: Myers, G. S. (1935): Four new freshwater fishes from Brasil, Venezuela and Paraguay. *Proc. Biol. Soc., Washington,* vol. 48, p. 11–13 (*Taeniacara candidi*).

Distribution: Upper Rio Negro region in Brazil.

First importation into Germany: 1976.

Characters:
fins: D XVI/6, A III/6, P 11

total length: male up to 70 millimeters; female up to 50 millimeters.

Sex differences: Males are conspicuously more colorful with a lanceolate, beautifully-colored tail fin tapered to a point as well as strikingly elongated ventral fins. Females are plainly colored and have a rounded-off tail fin.

Aquarium keeping: These fish are best kept in pairs or one male together with several females. On account of the small body size of this fish species, one can use smaller aquaria, which nevertheless should be at least 70 centimeters long. If several females are kept one should choose a larger tank. As a substrate, fine-grained gravel with a grain size of up to 3 millimeters is used. Decorating with resinous

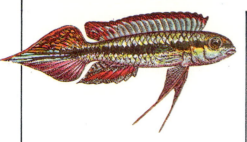

bog wood and a substantial planting contribute to the well-being of the fish. Rock formations or halved coconut shells serve as hiding places and spawning caves. In the furnishing of the aquarium it should be kept in mind that females set up territories around their cave. The distance between the caves therefore should be at least 30 centimeters.

In contrast to the majority of other species of dwarf cichlids, based on my previous experience it is necessary to offer this species soft and above all carbonate-poor water even for keeping. In addition, frequent water changes are advisable. The water temperature can be between 25 and 28° C. The fish should be fed principally with *Cyclops*.

It is not advisable to keep several males in small tanks since these fight fiercely and, therefore, losses cannot be ruled out. One can keep smaller tetras, such as *Petitella georgiae*, in the same tank without further ceremony.

Spawning: In this connection it is particularly important to use soft water, the total hardness of which should be less than 6° dH and the carbonate hardness less than 1° dH. The pH can be between 5 and 6. The water is best filtered through a peat filter.

The courting male swims diagonally with head down and with spread fins around the female, in the course of which it exhibits an up-and-down movement. In the female as well one can view this movement as a sign of her readiness to spawn. In this manner the females entice the males into the cave. The female attaches the eggs to the wall of the cave, where they are fertilized by the male. The clutch is then

exclusively tended by the female. After about two days the female lends assistance in the hatching of the fry. After a further four days the fry are free-swimming and must be fed with nauplia. The fry are cared for by the female. The further rearing is not difficult. It has proved effective to remove the clutch one day after spawning and to rear it artificially to prevent the female from eating the spawn.

Genus *Teleogramma* Boulenger, 1899

Explanation of the scientific name: Refers to the continuous, readily visible lateral line.
teleios (Gk.) = complete; *gramma* (Gk.) = mark, writing.

Original description: Boulenger, G. A. (1899): Matériaux pour la faune du Congo. Poissons nouveaux du Congo. *Ann. Mus. Congo, Zool.,* vol. 1, p. 53.

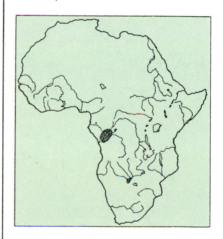

Distribution: Congo (Zaire) River basin, in fast-flowing and relatively shallow places.

Number of species: So far four species are known.

Generic characters: Fishes with elongated, cylindrical build, the swim bladder of which is no longer functional. The fish chiefly move themselves about by hopping forward abruptly. Undivided lateral line present, continuing as far as the tail fin. Dorsal fin with 24 spines; anal fin with five spines.

Total length: Up to 130 millimeters.

Type species: *Teleogramma gracile* Boulenger, 1899.

Comments: So far only *Teleogramma brichardi* has been imported alive and kept in aquaria.

Teleogramma brichardi
Poll, 1959
Brichard's Slender Cichlid

Explanation of the scientific name: Refers to the Belgian collector and author, Pierre Brichard (Burundi).

Other name in the literature: *Teleogramma gracile* Ladiges, 1958.

Original description: Poll, M. (1959): Recherches sur la faune ichthyologique de la région du Stanley Pool. *Ann. Mus. Royal, Congo Belge, Sci. Zool.,* vol. 71, p. 75–174 (*Teleogramma brichardi*).

Distribution: Congo (Zaire) River basin from Stanley Pool to Matadi.

Habitat: Fast-flowing, shallow bodies of water with a substrate of pebbles. Water with a hardness of less than 2° dH and a pH of about 7 at water temperatures of between 25° and 28° C, depending on season.

First importation into Germany: 1957.

Characters:
fins: D XXIV/7, A V/10
scales: mLR 60
total length: male up to 120 millimeters; female up to 100 millimeters.

Sex differences: Males are more slender and are almost totally black in color; females have a clearly visible, broad white edging on the dorsal fin and in the upper part of the tail fin. In addition, females exhibit a red belly area, particularly when in breeding condition. The red color

Steatocranus casuarius.

Steatocranus tinanti.

Taeniacara candidi ♂.

Teleogramma brichardi, ♂ in front of the ♀.

Teleogramma brichardi ♀, courtship coloration.

Teleogramma brichardi is a goby-like cichlid which stays close to the bottom most of the time because their swim bladders have become functionless. In the aquarium they stay hidden under rocks and in caves and, unless they are supplied with sufficient hiding places, the fishes do not thrive and perish very soon thereafter. On the other hand, they require a substantial amount of fresh water because of their unusually high oxygen requirement. This can only be achieved with a turbulent aquarium current. This is definitely not a fish for beginners, but presents an interesting challenge to serious fish breeders.

can extend right into the dorsal fin. The belly area then is also clearly fuller.

Aquarium keeping: Since Brichard's Slender Cichlid chiefly stays hidden under rocks or in shallow caves under natural conditions and comes together only briefly at spawning time, we must keep this in mind in the selection and furnishing of the keeping tank. Based on my own experiences, it is fitting to use large, above all long tanks at least 100 centimeters long. As a substrate, gravel with a grain size of up to 3 millimeters is suitable. Before one introduces the gravel, however, the rock structures are first set up directly on the bottom of the tank. Using flat rocks at fairly large intervals, several caves are built that are not too high, but instead as wide as possible. After the gravel has been put in, one plants the tank, but not in the immediate vicinity of the entrances of the caves, since the fish dig heavily there.

One can use virtually any tap water for keeping these fish. The water temperature should be between 23° and 28° C. The use of a powerful filter pump has proved effective in order to achieve a current as well as supplementary enrichment of oxygen. It is best to feed with midge larvae or small shrimp, particularly when preparing for spawning. Of course, any of the other usual live and dry foods are also eaten. At first we do not see all too much of the fish, since they almost exclusively stay in their holes. After they have overcome their initial shyness, it is possible to observe them outside of their caves more frequently. A few additional fishes, such as African tetras, help this fish to overcome its shyness more quickly.

Spawning: If one observes a considerable swelling of the female's belly area and a change in its coloration to red, if it courts frequently with curved body in front of the male and tries to entice the male into the cavity, one performs a partial water change. Half of the water is siphoned out. After that one allows lukewarm, totally desalinated water to run slowly into the tank through a thin hose. Filtering over peat is also recommended. Under these conditions one can expect the fish to now begin spawning, and for the 20 to 35 eggs attached to the roof of the cave to be fertilized and develop into fry. The yellowish-white eggs are about 3 millimeters in diameter. After spawning, the clutch is cared for by the female. The female often isolates the larvae from the environment until they hatch (depending on water temperature, three to four days after spawning), in that it closes off the entrance of the cave with sand. After this time the female occasionally comes out of the cave to feed. The fry, however, do not leave the cave for the first time until about twelve days later, and are then barely cared for by the parents anymore. They restrict themselves to territorial defense.

Genus *Telmatochromis* Boulenger, 1898

Explanation of the scientific name: Refers to the so-called canine teeth and to the former name for cichlids, "chromides."
telum (L.) = weapon; *chroma* (Gk.) = color.
Original description: Boulenger, G. A. (1898): Report on the collection of fishes made by Mr. J. E. S. Moore in Lake Tanganyika . . . 1895–1896. *Trans. Zool. Soc., London*, vol. 15, p. 10.
Distribution: Lake Tanganyika (Africa).

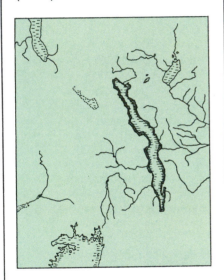

Number of species: So far five species are known.
Generic characters: Elongated body, two incomplete lateral lines, mouth heavily toothed, head blunt in old age, dorsal fin with 18 to 24 spines, anal fin with eight to nine spines.
Total length: 70 to 120 millimeters.
Type species: *Telmatochromis vittatus* Boulenger, 1898.

Telmatochromis bifrenatus Myers, 1936
Two-BandedTanganyika Cichlid

Explanation of the scientific name: Refers to the stripes of the fish.
bi- (L.) = twice, double; *frenum* (L.) = bridle.
Original description: Myers, G. S. (1936): Report on the fishes collected by H. C. Raven in Lake Tanganyika in 1920. *Proc. U.S. Nat. Mus.*, 84, p. 1–5 (*Telmatochromis bifrenatus*).
Distribution: Lake Tanganyika at Kigoma (Tanzania).
Habitat: Rocky shores at depths of 5 to 20 meters.
First importation into Germany: 1972.
Characters:
fins: D XXI-XXII/8, A VII/6
total length: male up to 90 millimeters; female up to 75 millimeters.
Sex differences: No external characters.
Aquarium keeping: As in *Julidochromis dickfeldi*.
Spawning: As in *Julidochromis dickfeldi*; however, brood care is carried out by the female considerably more intensively.

Telmatochromis caninus Poll, 1942

Explanation of the scientific name: Refers to the appearance of the head.
canis (L.) = dog; *caninus* (L.) = like a dog.
Other name in the literature: *Lamprologus dhouti*?
Original description: Poll, M. (1942): *Rev. Zool. Bot. Afr.,* iss. 53, p. 324 (*Telmatochromis caninus*).
Distribution: Lake Tanganyika.
Habitat: Rocky shores at depths of up to 60 meters.
First importation into Germany: 1970.

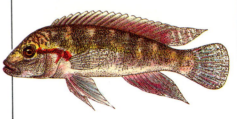

Characters:
fins: D XVIII-XX/7–10, A V-VIII/5–8
scales: mLR 31–36, strongly toothed
total length: male up to 120 millimeters; female up to 100 millimeters.
Sex differences: Males have a blunt head.
Aquarium keeping: As in *Julidochromis dickfeldi*, although a tank at least 100 centimeters long should be used.

A pair of *Telmatochromis caninus* spawning.

Spawning: As in *Julidochromis dickfeldi*, although a substantially larger number of eggs (up to 500) are laid. The clutch and fry are intensively cared for and guarded principally by the female.

Telmatochromis temporalis
Boulenger, 1898

Explanation of the scientific name: Refers to the dark stripe from the eye to the edge of the gill cover that can be clearly seen only in fright coloration.
temporalis (L.) = for a time, transitory.

Original description: Boulenger, G. A. (1898): Report on the collection of fishes made by Mr. J. E. S. Moore in Lake Tanganyika . . . 1895–1896. *Trans. Zool. Soc., London,* vol. 15, p. 11 (*Telmatochromis temporalis*).

Distribution: Lake Tanganyika (Kinyamkolo, Mbity Rocks, Kibwesa).

Habitat: Boulder and rocky shores at a depth of up to 8 meters.

First importation into Germany: 1956.

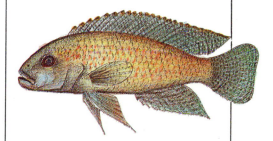

Characters:
fins: D XVIII-XXI/6–8, A VI-VII/6–7

total length: male up to 110 millimeters; female up to 85 millimeters.

Sex differences: Males have a forehead hump in old age.

Aquarium keeping: As in *Telmatochromis caninus*, although larger tanks should be used.

Spawning: As in *Telmatochromis caninus* (up to 400 eggs).

This is the fish which is sold in the aquarium trade as *Telmatochromis dhouti*, but the present author prefers to treat this as a synonym for *T. caninus*. The *caninus* name is derived from the canine-like teeth easily seen in the accompanying photograph.

Telmatochromis vittatus.

Telmatochromis caninus.

Telmatochromis temporalis.

Melanochromis johannii.

Oreochromis (Alcolapia) alcalicus grahami.

Eretmodus cyanostictus.

Labidochromis textilis.

Telmatochromis bifrenatus.

Variabilichromis moorei.

Telmatochromis vittatus
Boulenger, 1898

Explanation of the scientific name: Refers to the markings of this fish.

vittatus (L.) = striped.

Original description: Boulenger, G. A. (1898): Report on the collection of fishes made by Mr. J. E. S. Moore in Lake Tanganyika . . . 1895–1896. *Trans. Zool. Soc., London,* vol. 15, p. 10 (*Telmatochromis vittatus*).

Distribution: Lake Tanganyika (Mbity Rocks).

Habitat: Boulder and rocky shores, at a depth of up to 10 meters.

First importation into Germany: 1973.

Characters:

fins: D XXI-XXII/8, A XII/6–8

total length: both sexes up to 80 millimeters.

Sex differences: No external characters.

Aquarium keeping: As in *Julidochromis dickfeldi.*

Spawning: As in *Julidochromis dickfeldi,* although a larger number of eggs (up to 200) are laid.

Genus Variabilichromis
Colombe and Allgayer, 1985

Explanation of the scientific name: Refers to the color change of the fish during the development of the fry to the sexually mature fish as well as to the variability in the bone structure from individual to individual and to the former name for cichlids, "chromides."

variabilis (L.) = variable; *chroma* (Gk.) = color.

Original description: Colombe, J., and Allgayer, R. (1985): Description de *Variabilichromis, Neolamprologus* et *Paleolamprologus* genres nouveaux du Lac Tanganyika, avec redescription des genres *Lamprologus* Schilthuis, 1891 et *Lepidiolamprologus* Pellegrin, 1904 (Pisces: Teleostei: Cichlidae). *Rev. Franc. Cichl.,* nr. 5, p. 9–16 and 21–28.

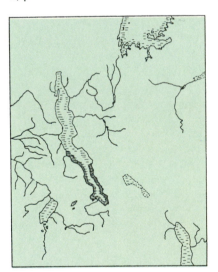

Distribution: Lake Tanganyika (Africa).

Number of species: So far only one species is known.

Generic characters: Differs from the genus *Neolamprologus* on the basis of the variable bone structure from fish to fish.

Total length: 100 millimeters.

Type species: *Variabilichromis moorei* (*Lamprologus moorei* Boulenger, 1898).

Variabilichromis moorei
(Boulenger, 1898)

Explanation of the scientific name: Refers to Prof. J. E. S. Moore.

Other name in the literature: *Lamprologus moorei* Boulenger, 1898.

Original description: Boulenger, G. A. (1898): Report

on the collection of fishes made by Mr. J. E. X. Moore in Lake Tanganyika . . . 1895–1896. *Trans. Zool. Soc., London,* vol. 15, p. 8 (*Lamprologus moorei*).

Distribution: Lake Tanganyika.

Habitat: Boulder and rock zones, at a depth of up to 3 meters.

First importation into Germany: 1975.

Characters:

fins: D XIX-XX/8–9, A VII/VIII/6–7

scales: mLR 33–35

total length: both sexes up to 100 millimeters.

Sex differences: No external characters are visible.

Aquarium keeping: As in *Neolamprologus buescheri.*

Spawning: As in *Neolamprologus buescheri.*

Genus *Chromidotilapia* Boulenger, 1898

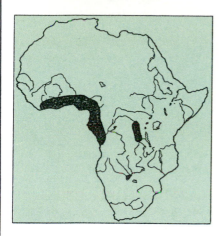

Explanation of the scientific name: Refers to the coloration of these fishes.

chroma (Gk.) = color; *Tilapia* = genus of African cichlids.

Original description: Boulenger, G. A. (1898): A revision of the African and Syrian fishes of the family Cichlidae. *Proc. Zool. Soc., London,* p. 151.

Distribution: Western Africa from Sierra Leone to the Congo.

Number of species: So far nine species are known.

Generic characters: Both sexes have virtually equally large ventral fins and clearly visible nipple-like protrusions on either side of the roof of the gullet.

Total length: 100 to 200 millimeters.

Type species: *Chromidotilapia kingsleyae* Boulenger, 1898.

Comments: The species described before 1968 were classified in the genus *Pelmatochromis*. After the revision by Thys van den Audenaerde (1968), the genus *Chromidotilapia* was revived as a subgenus and today once again has generic status.

Chromidotilapia finleyi Trewavas, 1974

Explanation of the scientific name: Refers to L. Finley.

Other names in the literature: *Pelmatochromis nigrofasciatus* Boulenger, 1915
Pelmatochromis batesii Thys van den Audenaerde, 1967.

Original description: Trewavas, E. (1974): The freshwater fishes of rivers Mungo and Meme and Lake Kotto, Mboandong and Soden, west Cameroon. *Bull. British Mus. Nat. Hist. (Zool.),* 26, p. 393–397 (*Chromidotilapia finleyi*).

Distribution: After Linke, the Mungo River and its northern tributaries as well as the Lobe river basin in western Cameroon.

Habitat: Small water courses with clear, brownish water, partially with growth of *Bolbitis heudelotti* and various *Anubias* species. Hardness less than 1° dH, in part with a *p*H of about 5, in other places also a *p*H of 7.6 (Linke and Staeck, 1981). Water temperature about 24° C.

First importation into Germany: 1973 by H. Linke (West Berlin).

Characters:
fins: D XV–XVI/9–10, A III/7–8
scales: mLR 26–28
total length: male up to 120 millimeters; female up to 100 millimeters.

Sex differences: Males are more slender. Females with fuller and more reddish belly area.

Aquarium keeping: These large fish are best kept in pairs in a fairly large tank with a length of at least 100 centimeters. Sufficient good hiding places in the form of cavities, rock structures, and resinous bog roots must be available. The rock structures are put directly on the bottom glass, since otherwise they could collapse when the fish dig. As a substrate one uses gravel with a grain size of up to 4 millimeters. The best-suited plants are the robust *Anubias* species and *Bolbitis heudelotti*, which one also anchors with rocks in addition to inserting them in the substrate. Normal tap water with a temperature of about 25° C is quite sufficient for keeping. One can also keep several African tetras together with the pair.

Spawning: The fish can stay in the keeping tank, although with fairly hard tap water it is better to replace half of the tank's water with totally desalinated water before the fish spawn. As a rule, the fish spawn on a rock that is cleaned by the female before spawning. Usually the female takes the eggs into her mouth in the last phase of spawning. In some cases they are also taken up by the male. One often observes that the male excavates a hollow near the spawning site before spawning begins. The female places the eggs that are already in her mouth into this hollow. They are then perhaps taken immediately or somewhat later by the male into his mouth. Later this also occurs in reverse form when the female takes the eggs from the male. The fish thus alternate in brooding the eggs in their mouths. The larvae hatch after barely five days and then develop in the parents' mouths into fry in the following six to seven days. Then they are let out of the mouth. Both mates also take part in the further brood care.

Comments: Various color varieties of this fish species have been collected.

Chromidotilapia finleyi.

Chromidotilapia linkei.

Cyprichromis leptosoma with yellow tail fin.

Cyprichromis leptosoma with blue tail fin.

Cyprichromis nigripinnis.

Chromidotilapia linkei
Staeck, 1980

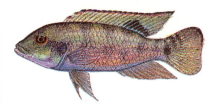

Explanation of the scientific name: Refers to H. Linke (West Berlin).

Original description: Staeck, W. (1980): *Chromidotilapia linkei* n. sp. aus dem Mungo River, Kamerun (Pisces: Cichlidae). *Senckenbergiana biol.* 60(3/4), p. 153–157 (*Chromidotilapia linkei*).

Distribution: Mungo River system (Cameroon).

Habitat: Shallow water zones, among branches and under overhanging riparian vegetation; clear water with a total hardness of about 3° dH and a *p*H of between 7.6 and 7.8; water temperature about 26° C.

First importation into Germany: 1977 by H. Linke (West Berlin).

Characters:
fins: D XV/10–11, A III/7
scales: mLR 25–26
total length: male up to 100 millimeters; female up to 90 millimeters.

Sex differences: Males are more slender; females have a light, metallic-colored stripe on the dorsal fin.

Aquarium keeping: As in *Chromidotilapia finleyi.*

Spawning: As in *Chromidotilapia finleyi.*

Chromidotilapia schoutedeni
(Poll and Thys, 1967)

Explanation of the scientific name: Refers to Dr. H. Schouteden, honorary director of the Musée Royal de l'Afrique centrale, Tervuren.

Other names in the literature:
Pelmatochromis longirostris Poll, 1954
Pelmatochromis schoutedeni Poll and Thys van den Audenaerde, 1967.

Original description: Poll, M., and Thys van den Audenaerde, D. F. E. (1967): Description de *Pelmatochromis schoutedeni* sp. n. du Congo oriental (Pisces: Cichlidae). *Rev. Zool. Bot. Afr.,* vol. 75 (3–4), p. 383–391 (*Pelmatochromis schoutedeni*).

Distribution: Upper Congo (Zaire) River system, Lomboma River, Lubilu River, at Shabunda and Kinangani (Stanleyville) in Zaire.

First importation into Germany: 1967.

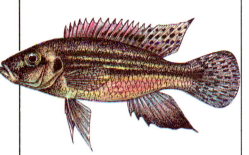

Characters:
fins: D XIV-XV/9–10, A III/7–8
scales: mLR 18–21
total length: male up to 90 millimeters; female up to 80 millimeters.

Sex differences: Males are more slender; females have a light metallic-colored band in the dorsal fin.

Aquarium keeping: As in *Chromidotilapia finleyi.*

Spawning: As in *Chromidotilapia finleyi.*

Comments: Differing from the females of other species of this genus, the females of this species possess rounded-off ventral fins.

Genus *Cyprichromis*
Scheuermann, 1977

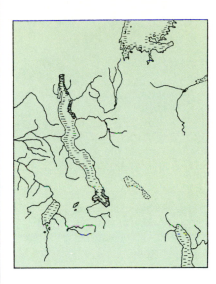

Explanation of the scientific name: Refers to the build of the fishes and to the former name for cichlids, "chromides."
Cyprinoidea = suborder of carplike fishes; *chroma* (Gk.) = color.

Original description: Scheuermann, H. (1977): A partial revision of the genus *Limnochromis* Regan, 1920. *Cichlidae,* iss. 3, nr. 2, p. 69–73 (Belge).

Distribution: Lake Tanganyika (Africa).

Number of species: So far three species are known.

Generic characters: School fish with a herringlike appearance.

Total length: 100 to 140 millimeters.

Type species: *Cyprichromis leptosoma* (*Paratilapia leptosoma* Boulenger, 1898).

Comments: This genus was taken out of the genus *Limnochromis* by Scheuermann (1977).

Cyprichromis leptosoma
(Boulenger, 1898)

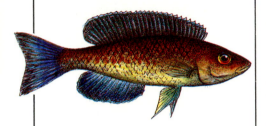

Explanation of the scientific name: Refers to the build of the fish.

leptos (Gk.) = slender; *soma* (Gk.) = body, trunk.

Other names in the literature:
Paratilapia leptosoma
Boulenger, 1898
Limnochromis leptosoma
Regan, 1920.

Original description:
Boulenger, G. A. (1898): Report on the collection of fishes made by Mr. J. E. S. Moore in Lake Tanganyika . . . 1895–1896. *Trans. Zool. Soc., London,* vol. 15, p. 14 (*Paratilapia leptosoma*)

Distribution: Lake Tanganyika (Mbete, Kinyamkolo, Msambu, Kigoma).

Habitat: Open water at depths below 5 meters; on rocky shores in large schools.

First importation into Germany: 1975.

Characters:
fins: D XII-XIV/14–16, A III/10–12
scales: mLR 39–42
total length: male up to 110 millimeters; female up to 100 millimeters.

Sex differences: Males are more colorful.

Aquarium keeping: It has proved to be beneficial to keep this species in a school. One should put at least eight fish in the tank. This of course demands a correspondingly large tank with a capacity of about 150 liters and a length of at least 130 centimeters. As decoration one uses fairly large rocks and aquatic plants. In the foreground of the tank ample open swimming space should be offered, but one can, however,

plant small-growing *Anubias*. Medium-hard to hard water with a *p*H of not less than 7 should be used. The water temperature can be between 23 and 28° C.

Spawning: The fish spawn in open water. In the aquarium, courtship and spawning are generally observed just under the water's surface. The released eggs are immediately taken into the female's mouth along with the male's sperm, which is readily visible in the water. The fry are let out of the mouth after 23 to 28 days. They immediately gather into a school at the water's surface. One feeds them with *Artemia* or *Cyclops* nauplia.

The chief requirement for successful spawning is the use of the largest possible tank.

Comments: Various color varieties were collected from the large schools of this fish species; those possessing a yellow to orange colored as well as a violet-blue tail fin are particularly striking.

Cyprichromis microlepidotus
(Poll, 1956)

Explanation of the scientific name: Refers to the size of the scales.

micros (Gk.) = small; *lepis* (Gk.) = scale.

Other name in the literature:
Limnochromis microlepidotus
Poll, 1956.

Original description: Poll, M. (1956): Poissons Cichlidae. Résultats scientifiques exploration hydrobiologique du Lac Tanganyika 1946–1947. vol. 3, nr. 5B, p. 185–187 (*Limnochromis microlepidotus*).

Distribution: Lake Tanganyika.

First importation into Germany: 1975.

Characters:
fins: D XII-XIV/15–18, A III/12–13
total length: male up to 110 millimeters; female up to 100 millimeters.

Sex differences: Males, as in *Cyprichromis leptosoma*, are conspicuously colored, particularly at spawning time. Females are a plain yellow-gray.

Aquarium keeping: As in *Cyprichromis leptosoma*.

Spawning: As in *Cyprichromis leptosoma*.

Cyprichromis nigripinnis
(Boulenger, 1901)

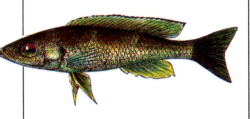

Explanation of the scientific name: Refers to the coloration of the fins.

niger (L.) = black, blackish; *pinna* (L.) = fin.

Other names in the literature:
Paratilapia nigripinnis
Boulenger, 1901
Limnochromis nigripinnis
Regan, 1920.

Original description:
Boulenger, G. A. (1901): *Ann. Mag. Nat. Hist.,* ser. 7, vol. 7, p. 3 (*Paratilapia nigripinnis*).

Distribution: Lake Tanganyika (Msambu).

Habitat: Open water at depths of greater than 10 meters; in large schools.

First importation into Germany: 1975.

Characters:
fins: D XV-XVII/11, A III/8–9
scales: mLR 39–40
total length: male up to 110 millimeters; female up to 100 millimeters.

Sex differences: Males have a yellow edging to the dorsal fin and a yellow coloration to the soft-rayed part of the dorsal fin. Females have a yellow coloration in the ventral fins and in the anal fin.

Aquarium keeping: As in *Cyprichromis leptosoma*.

Spawning: As in *Cyprichromis leptosoma*.

Genus *Eretmodus* Boulenger, 1898

Explanation of the scientific name: Refers to the form of the ventral fins.

eretmon (Gk.) = rudder.

Original description: Boulenger, G. A. (1898): Report on the collection of fishes made by Mr. J. E. S. Moore in Lake Tanganyika . . . 1895–1896. *Trans. Zool. Soc., London,* vol. 15, p. 16.

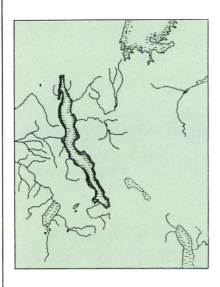

Distribution: Lake Tanganyika (Africa).

Number of species: So far only one species is known.

Generic characters: Cichlids that have adapted to life on the bottom. The swim bladder is vestigial, so that the fish are no longer able to swim freely in the water. Dorsal fin with 22 to 25 spines; anal fin with three spines.

Total length: Up to 90 millimeters.

Type species: *Eretmodus cyanostictus* Boulenger, 1898.

Eretmodus cyanostictus, head study.

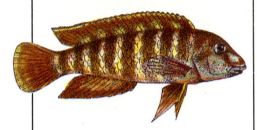

Eretmodus cyanostictus
Boulenger, 1898
Tanganyika Clown Cichlid

Explanation of the scientific name: Refers to the iridescent bluish spots that are present, particularly in the head region.
kyaneos (Gk.) = blue; *stigma* (Gk.) = mark, brand, mark on the forehead.

Original description: Boulenger, G. A. (1898): Report on the collection of fishes made by Mr. J. E. S. Moore in Lake Tanganyika . . . 1895–1896. *Trans. Zool. Soc., London,* vol. 15, p. 16–30 (*Eretmodus cyanostictus*).

Distribution: Lake Tanganyika (Mbity Rocks, Kinyamkolo, Niamkolo Bay, Sumba, Kigoma).

Habitat: Shallow boulder zones up to very close to shore.

First importation into Germany: 1958.

Characters:
fins: D XXIII-XXV/3–5, A III/6–7
scales: mLR 32–35
total length: male up to 90 millimeters; female up to 80 millimeters.

Sex differences: Males are somewhat more slender; otherwise, scarcely any external characters.

Aquarium keeping: Fairly large tanks with rock structures and thick planting in the background are the most suitable for this species, which should be kept in pairs if possible. It is best to wait for a pair to form from a group of about six juveniles and then to remove the remaining fish from the tank. For feeding, vegetable food, such as algae or scalded lettuce, should be offered. Appropriate dry food (high vegetable content) is also suitable.

Spawning: No special preparations are necessary. Ready-to-spawn fish clean the spawning site only briefly and begin circling immediately afterwards. At the same time the mates nudge one another in the anal region. Then the first egg is released and is immediately taken into the female's mouth. With a diameter of about 5 millimeters, the eggs are very large. Subsequently, the female again touches the male's ventral side with her mouth as he presents it to her in a diagonal position, in the course of which she probably takes up sperm. Fertilization apparently occurs in the female's mouth. After spawning, the pair continues to remain in contact, contrary to the habits of other mouthbrooders. About twelve days after spawning, the male receives the larvae after a transfer ritual and does not release them until they have completed their development into fry.

Comments: Because of its comical appearance and clownish courtship behavior, this fish is considered as the clown among cichlids.

**Genus *Iodotropheus*
Oliver and Loiselle, 1972**

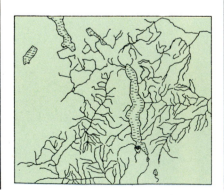

Explanation of the scientific name: Refers to the russet head and back area o the type material.
ios (Gk.) = rust; *Tropheus* = cichlid genus of Lake Tanganyika, established by Boulenger in 1898.

Original description: Oliver, M. K., and Loiselle, P. V. (1972): A new genus and species of Cichlid of the Mbuna group from Lake Malawi. *Rev. Zool. Bot. Afr.,* vol. 85, p. 309–320.

Distribution: Lake Malawi (Nyasa, Africa).

Number of species: So far only one species is known.

Generic characters: Build resembles that of the genus *Tropheus*; the egg spots are bordered in black.

Total length: Up to 100 millimeters.

Type species: *Iodotropheus sprengerae*
Oliver and Loiselle, 1972.

Iodotropheus sprengerae Oliver and Loiselle, 1972

Explanation of the scientific name: Refers to the proper name Sprenger.

Original description: Oliver, M. K., and Loiselle, P. V. (1972): A new genus and species of Cichlid of the Mbuna group from Lake Malawi. *Rev. Zool. Bot. Afr.,* vol. 85, p. 309–320 (*Iodotropheus sprengerae*).

Distribution: Southeastern Lake Malawi in the region of the Boadzulu Islands.

Habitat: Rock and boulder zone with heavy algal growth.

First importation into Germany: 1978.

Characters:
fins: D XVI-XVIII/6–8, A V/6
total length: male up to 100

millimeters; female up to 80 millimeters.

Sex differences: The male exhibits an intense blue to violet coloration of the sides of the body; on the anal fin are found up to five black-bordered egg spots. The female has a maximum of only three egg spots, which also are not as prominently bordered as in the male. Females are red-brown colored.

Aquarium keeping: As in *Melanochromis auratus*.

Spawning: As in *Melanochromis auratus*.

Genus *Labidochromis* Trewavas, 1935

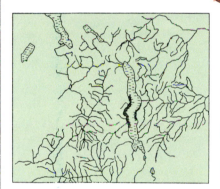

Explanation of the scientific name: Refers to the partially elongated unicuspid teeth and to the former name for cichlids, "chromides."

labis (Gk.) = tongs; *chroma* (Gk.) = color.

Original description: Trewavas, E. (1935): A synopsis of the Cichlid fishes of Lake Nyasa. *Ann. Mag. Nat. Hist.,* ser. 10, vol. 16, p. 69, 80 and 117.

Distribution: Lake Malawi.

Number of species: So far eight species are known.

Generic characters: Both sexes have egg spots.. Externally, *Labidochromis* species can scarcely be distinguished from *Pseudotropheus* species. Differences exist in the dentition: the former have an outer row of unicuspid teeth directed forwards.

Total length: 80 to 100 millimeters.

Type species: *Labidochromis vellicans* Trewavas, 1935.

Labidochromis caeruleus Fryer, 1956

Explanation of the scientific name: Refers to the body coloration.

caeruleus (L.) = blue, bluish.

Original description: Fryer, G. (1956): New species of Cichlid fishes from Lake Nyasa. *Rev. Zool. Bot. Afr.,* vol. 53, p. 81–91 (*Labidochromis caeruleus*).

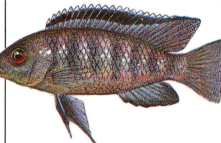

Distribution: Lake Malawi (Nkata Bay).

Habitat: Rocky bottom, but also in *Vallisneria* beds.

First importation into Germany: 1958.

Characters:
fins: D XVIII/9, A III/7
total length: male up to 100 millimeters; female up to 80 millimeters.

Sex differences: No external characters are visible.

Aquarium keeping: As in *Melanochromis auratus*.

Spawning: As in *Melanochromis auratus*.

Labidochromis freibergi Johnson, 1974

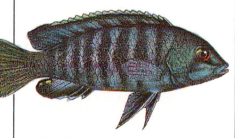

Explanation of the scientific name: Refers to the proper name Freiberg.

Other name in the literature:

Labidochromis ewarti?

Original description: Johnson, D. S. (1974): Three new Cichlids from Lake Malawi. *Today's Aquarist,* 1, p. 38–42 (*Labidochromis freibergi*).

Distribution: Lake Malawi, at the islands of Chisumulu and Likoma.

First importation into Germany: 1964.

Characters:
fins: D XVIII/7–8, A III/7
total length: male up to 80 millimeters; female up to 70 millimeters.

Sex differences: No external characters are visible.

Aquarium keeping: As in *Melanochromis auratus*.

Spawning: As in *Melanochromis auratus*.

Labidochromis mathotho Burgess and Axelrod, 1976

Explanation of the scientific name: Refers to A. Mathoto, the Chief Fisheries Officer of Malawi.

Original description: Burgess, W. E., and Axelrod, H. R. (1976): Two new species of Mbuna (Rock-Dwelling Cichlids) from Lake Malawi. *Tropical Fish Hobbyist,* vol. 24, nr. 7, p. 48–52 (*Labidochromis mathotho*).

Distribution: Lake Malawi.

Characters:
fins: D XVI-XVII/8–9, A III/7, P 13
scales: mLR 29–30
total length: male up to 80 millimeters; female up to 75 millimeters.

Sex differences: No external characters.

Aquarium keeping: As in *Melanochromis auratus*.

Spawning: As in *Melanochromis auratus*.

Labidochromis textilis Oliver, 1974

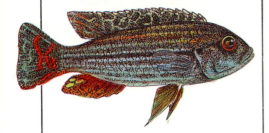

Explanation of the scientific name: Refers to the pattern of markings, which resembles a fabric.

textilis (L.) = fabric.

Original description: Oliver, M. K. (1974): *Labidochromis textilis*, a new Cichlid fish (Teleostei: Cichlidae) from Lake Malawi. *Proc. Biol. Soc. Washington,* vol. 88, p. 319–330 (*Labidochromis textilis*).

Distribution: Lake Malawi.

First importation into Germany: 1978.

Characters:
fins: D XI/9–10, A III/7–8
scales: mLR 28–36
total length: male up to 75 millimeters; female up to 70 millimeters.

Sex differences: No external characters.

Aquarium keeping: As in *Melanochromis auratus.*

Spawning: As in *Melanochromis auratus.*

Labidochromis vellicans Trewavas, 1935

Pointed-Head Mouthbrooder

Explanation of the scientific name: Refers to the typical manner of food intake.

vellicere (L.) = to pluck.

Original description: Trewavas, E. (1935): A synopsis of the Cichlid fishes of Lake Nyasa. Ann. Mag. Nat. Hist., ser. 10, vol. 16, p. 80 (*Labidochromis vellicans*).

Distribution: Lake Malawi (Monkey Bay, Nkudzi Bay).

Habitat: Rocky and sandy shores.

First importation into Germany: 1958.

Characters:
fins: D XV-XVIII/9–10, A III/7–8
scales: mLR 30–32
total length: male up to 100 millimeters; female up to 90 millimeters.

Sex differences: Males have small orange-red egg spots and a bright-blue body coloration that are lacking in the female.

Aquarium keeping: As in *Melanochromis auratus.*

Spawning: As in *Melanochromis auratus.*

Genus *Melanochromis* Trewavas, 1935

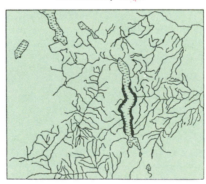

Explanation of the scientific name: Refers to the coloration of the majority of the males of this genus and to the former name for cichlids, "chromides."

melas (Gk.) = black, dark;
chroma (Gk.) = color.

Original description: Trewavas, E. (1935): A synopsis of Cichlid fishes of Lake Nyasa. *Ann. Mag. Nat. Hist.,* ser. 10, vol. 16, p. 77–78.

Distribution: Lake Malawi (Africa).

Number of species: So far nine species are known.

Generic characters: Adult males of the majority of species possess a dark ground color. The pharyngeal teeth are more strongly developed than in the *Pseudotropheus* species.

Total length: 80 to 150 millimeters.

Type species: *Melanochromis melanopterus* Trewavas, 1935.

Melanochromis auratus (Boulenger, 1897) Turquoise-Gold Cichlid

Explanation of the scientific name: Refers to the coloration of the fish.

auratus (L.) = gilded, golden, having gold.

Other names in the literature:
Chromis auratus Boulenger, 1897
Tilapia aurata Boulenger, 1898
Pseudotropheus auratus Regan, 1921.

Original description: Boulenger, G. A. (1897): *Ann. Mag. Nat. Hist.,* ser. 6, vol. 19, p. 155 (*Chromis auratus*).

Distribution: Lake Malawi (Monkey Bay).

Habitat: Rocky zones with algal growth.

First importation into Germany: 1958.

Characters:
fins: D XVIII-XIX/6–9, A III/6–8
scales: mLR 33–34
total length: male up to 110 millimeters; female up to 90 millimeters.

Sex differences: Adult males, when they have assumed their final coloration, are blue-black with two turquoise-colored longitudinal stripes, the middle one of which is particularly conspicuous. Females are golden-yellow and have two black longitudinal stripes.

Aquarium keeping: Fairly large aquaria (if possible over 100 centimeters long) with rock structures are suitable. In the foreground, fine-grained gravel

should be used as a substrate. No particular demands are made with respect to water quality. The *p*H should not fall below the neutral value of 7, but can range as high as 8. The water temperature should be about 25° C.

It has proved effective to keep several females with one male and to place other species of about the same body size with them in the tank. Then the aggression toward females feared in this species will occur less often.

Spawning: The male occupies a territory at spawning time and courts all females that swim by. Ready-to-spawn females swim to the male and, after the brief courtship behavior of the male, start to spawn at once. The released eggs are immediately taken into the female's mouth. After the sperm is taken in, fertilization takes place in the mouth. The female retreats to a protected site after spawning and cares for the brood through frequent shifting of the eggs in the mouth (chewing movements). The larvae stay in the female's mouth until they become free-swimming. This takes between 22 and 26 days. During this time, females are still able to take up food despite the full throat sack. The fry are about 10 millimeters long when they are released from the female's mouth, and in the following days are only occasionally taken back into the mouth. They seek out suitable hiding places very quickly, and then are no longer cared for by the female. The coloration of the fry resembles that of the female. When feasible, one should remove the fry from the tank immediately after they are released from the mouth and rear them separately. One can also catch the female with the full mouth and keep her for some time outside of the tank in the net. It will spit out the fry after a short time.

Melanochromis brevis
Trewavas, 1935
Red-Brown Mouthbrooder

Explanation of the scientific name: Refers to the compact build.

brevis (L.) = short.

Original description: Trewavas, E. (1935): A synopsis of the Cichlid fishes of Lake Nyasa. *Ann. Mag. Nat. Hist.,* ser. 10, vol. 16, p. 65–118 (*Melanochromis brevis*).

Distribution: Lake Malawi.

Habitat: Rocky zones with algal growth.

First importation into Germany: 1972.

Characters:

fins: D XVI/9, A III/6–8

scales: mLR 28–31

total length: male up to 120 millimeters; female up to 100 millimeters.

Sex differences: Males are more slender and darker, with a blue tint in the brownish coloration. The egg spots either are not present or are only suggested in the female.

Aquarium keeping: As in *Melanochromis auratus.*

Spawning: As in *Melanochromis auratus.*

Melanochromis exasperatus
(Burgess, 1976)
Pearl of Likoma

Explanation of the scientific name: Refers to the convoluted history of this species.

exasperatus (NL) = irritating.

Other names in the literature: *Labidiochromis caeruleus likomae?*

Pseudotropheus joanjohnsonae?

Original description: Burgess, W. E. 1976. Studies on the family Cichlidae: No. 4. Two new species of Mbuna (Rock-Dwelling Cichlids) from Lake Malawi. *Tropical Fish Hobbyist.* vol 24 (#241, No. 7) pp.44—52.

Distribution: Lake Malawi.

Habitat: Boulder zones of Likoma Island and the Chisumulu Islands.

First importation into Germany: 1972.

Characters:

fins: D XVII/6–7, A III/7–8

scales: mLR 28–32

total length: male up to 100 millimeters; female up to 80 millimeters.

Sex differences: Males are more slender and a rich blue, whereas females exhibit an iridescent blue-green ground color that is interrupted by gold-colored longitudinal bands.

Aquarium keeping: As in *Melanochromis auratus.* Feeding with insect larvae and vegetable food is appropriate.

Spawning: As in *Melanochromis auratus.*

Melanochromis johannii
(Eccles, 1973)
Cobalt-Orange Cichlid

Explanation of the scientific name: Refers to the fisherman Johan James.

Other names in the literature: *Pseudotropheus daviesi?*

This is a male *Melanochromis johanni*, also known incorrectly as *"Pseudotropheus" johanni*. This specimen does not show the usual light blue horizontal stripes. The female looks quite like another species as she is orange and not cobalt blue like the male! This sexual dimorphism has led to a great deal of confusion as breeders thought they were crossing two different species of fish.

Pseudotropheus johannii
Eccles, 1973.
Original description: Eccles, D. H. (1973): Two new species of Cichlid fishes from Lake Malawi. *Arnoldia, Rhodesia,* nr. 16, 6, p. 1–7 (*Pseudotropheus johannii*).
Distribution: Lake Malawi.
Habitat: Rock and boulder zones.
First importation into Germany: 1972.
Characters:
fins: D XVII/7, A III/8–9

scales: mLR 29–32
total length: male up to 100 millimeters; female up to 80 millimeters.
Sex differences: Males have a blue-black ground color on which are found two iridescent light blue longitudinal stripes. Females are gold-yellow colored.
Aquarium keeping: As in *Melanochromis auratus*.
Spawning: As in *Melanochromis auratus*.

Melanochromis perspicax
Trewavas, 1935
Explanation of the scientific name: Refers to the facial expression.
perspicax (L.) = sharp-sighted.
Original description: Trewavas, E. (1935): A synopsis of the Cichlid fishes of Lake Nyasa. *Ann. Mag. Nat. Hist.* ser. 10, vol. 16, p. 79 (*Melanochromis perspicax*).
Distribution: Lake Malawi (Deep Bay at Chilumba).

Pseudotropheus spec., blue.

Pseudotropheus spec., yellow.

Melanochromis exasperatus.

Melanochromis brevis.

Habitat: Rocky shore zones.
First importation into Germany: 1980.
Characters:
fins: D XVII/8, A III/6
scales: mLR 29–31
total length: male up to 100 millimeters; female up to 80 millimeters.

Sex differences: Males have a dark ground color, two blue-gray longitudinal stripes, and a light blue dorsal fin. Females have a whitish gray ground color and are easily confused with males of *Melanochromis parallelum*.
Aquarium keeping: As in *Melanochromis auratus*.
Spawning: As in *Melanochromis auratus*.

Genus *Oreochromis* Günther, 1889

Explanation of the scientific name: Refers to the cylindrical body and to the former name for cichlids, "chromides."
orca (L.) = barrel; *chroma* (Gk.) = color.
Original description: Günther, A. C. L. G. (1889): On some fishes from Kiloma-Njaro District. *Proc. Zool. Soc. London*, p. 70.

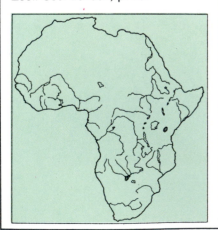

Subgenus *Alcolapia* Thys, 1968

Explanation of the subgeneric name: Refers to the occurrence of the fishes in alkaline saline lakes.
Original description: Thys van den Audenaerde, D. F. E. (1968): An annotated bibliography of *Tilapia* (Pisces: Cichlidae). *Doc. Zool. Mus. Royal de l'Afrique Centrale.* 14, p. 406.
Distribution: Saline lakes in East Africa (Lake Natron, Lake Magadi).
Number of species: So far eight species are known.
Subgeneric characters: Small-growing cichlid species from Lake Natron and Lake Magadi with almost cylindrical bodies.
Total length: Up to 100 millimeters.

Oreochromis (Alcolapia) alcalicus alcalicus (Hilgendorf, 1905)
Lake Natron Mouthbrooder

Explanation of the scientific name: Refers to the strongly alkaline water conditions.
alcalicus (L., Arabic) = basic.
Other names in the literature:
Tilapia alcalica Hilgendorf, 1905
Sarotherodon alcalicus Trewavas, 1972
Sarotherodon alcalicus alcalicus Trewavas, 1973.
Original description: Hilgendorf, F. M. (1905): *Zool. Jahrb. Abt. Syst.*, 22, p. 407 (*Tilapia alcalica*).
Distribution: Lake Natron and volcanic water holes near Lake Natron in Tanzania.
Habitat: Strongly alkaline water with a *p*H value of about 10.5. Muddy bottom with rocks covered with a thin layer of mud.
First importation into Germany: 1963.
Characters:
fins: D XIII/11, A III/11
total length: male up to 120 millimeters; female up to 110 millimeters.
Aquarium keeping: Tanks at least 70 centimeters long are needed. The fish are kept in pairs

or with other fishes in larger aquariums. They are somewhat pugnacious. The tank receives a substrate of fine-grained gravel, the background is well planted, and rocks of various sizes can serve as decoration.
The water hardness is insignificant; the *p*H, however, should not be less than the neutral value of 7. The water temperature should be 23° to 28° C.
Spawning: The fish spawn in a depression. In the process they turn in a circle in the manner of the majority of mouthbrooders. The female takes the about 1.5-millimeter-in-diameter yellow eggs into her mouth. A total of up to 30 eggs are laid. After about 24 days the female releases the fry from her mouth. These are then about 6 millimeters long and are fed with *Artemia* or *Cyclops* nauplia.

Oreochromis (Alcolapia) alcalicus grahami (Boulenger, 1912)
Magadi Mouthbrooder

Explanation of the scientific name: Refers to the collector, J. W. Graham.

Other names in the literature:
Tilapia grahami Boulenger, 1912
Sarotherodon grahami Trewavas, 1972
Sarotherodon alcalicus grahami Trewavas, 1973.
Original description: Boulenger, G. A. (1912): *Ann. Mag. Nat. Hist.* ser. 8, vol. 9, p. 519 (*Tilapia grahami*).
Distribution: Lake Magadi in Kenya.
Habitat: Strongly alkaline water with a *p*H of about 10.5. It was observed that the fish in the lake swam through areas with hot springs that exhibited a water temperature of about 40° C.
First importation into Germany: 1958.
Characters:
fins: D XI/11–12, A III/8–9
scales: mLR 28–30
total length: male up to 120 millimeters; female up to 100

Genus *Pseudocrenilabrus*

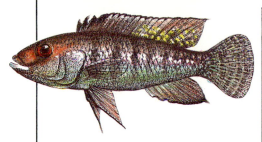

millimeters.

Sex differences: Males are more intensely colored.

Aquarium keeping: As in *Oreochromis (Alcolapia) alcalicus alcalicus.*

Spawning: As in *Oreochromis (Alcolapia) alcalicus alcalicus.*

Genus *Pseudocrenilabrus* Fowler, 1934

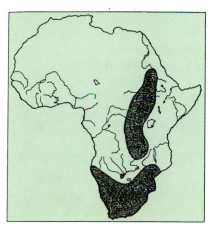

Explanation of the scientific name: Refers to the marine wrasse genus *Crenilabrus* Cuvier, 1815.

pseudo- (L.) = false; *crena* (L.) = notch; *labrum* (L.) = lip.

Original description: Fowler, H. W. (1934): *Pseudocrenilabrus* gen nov. *Proc. Acad. Nat. Sci., Phila.*, vol. 86, p. 462 f.

Distribution: Northeastern and southern Africa.

Number of species: So far three species are known.

Generic characters: Typical cichlid build with strikingly large head. Dorsal fin with 13 to 15 spines and 8 to 11 soft rays.

Total length: Up to 100 millimeters.

Type species: *Pseudocrenilabrus philander*

(Weber, 1897).

Comments: The species of the genus *Pseudocrenilabrus* originally belonged to the genus *Haplochromis*. Wickler (1963) created the genus *Hemihaplochromis* for the two species *Haplochromis multicolor* and *Haplochromis philander*. This generic name is invalid, however, since Fowler (1934) had already established the genus *Pseudocrenilabrus* for the species *Haplochromis philander*, which he nevertheless described for the second time under the name *Pseudocrenilabrus natalensis*. Hence, this generic name is valid. Trewavas (1973) pointed this out.

Pseudocrenilabrus multicolor (Hilgendorf, 1903) Multicolored Mouthbrooder

Explanation of the scientific name: Refers to the coloration of the fish.

multus (L.) = many, numerous; *color* (L.) = color.

Other names in the literature: *Paratilapia multicolor* Hilgendorf, 1903

Chromis multicolor Schoeller, 1903

Haplochromis strigigena (non Pfeffer) Boulenger, 1907

Haplochromis multicolor Regan, 1922

Hemihaplochromis multicolor Wickler, 1963.

Original description: Hilgendorf, F. M. (1903): *Sber. Ges. naturf. Freunde, Berlin, p. 429 (Paratilapia multicolor).*

Distribution: Northern Africa (upper Nile region to Tanzania).

Habitat: Places with heavy plant growth in marshy regions and small, slow-flowing bodies of water.

First importation into Germany: 1902 by Schoeller.

Characters:
fins: D XIII–XV/8–10, A III/6–9, P 12

scales: mLR 25–29

total length: male up to 70 millimeters; female up to 80 millimeters.

Sex differences: Males are more beautifully colored, particularly at spawning time. Females are yellow-gray.

Aquarium keeping: It is suitable to keep one male and two to four females in an approximately 70-centimeter-long tank. The tank is to be provided with a substrate of fine-grained gravel (about 2-millimeter grain size) and should be well planted in several places. No particular demands are made with respect to water hardness. The water temperature should be between 23° and 28° C.

Spawning: The male vigorously courts a ready-to-spawn female and in so doing displays its most beautiful coloration. A spawning depression is dug in the gravel by the female. The depression is smoothed out during the circling of the mates. During the circling the eggs are released into the hollow by the female and apparently are fertilized immediately by the male. It is assumed, however, that in this group of fishes as well the sperm is already sucked into the mouth by the female during the circling. The female always immediately takes up the eggs that are just behind the male's body.

After spawning, the female with the full throat sac retreats to a sheltered site. The fry are let out of the mouth after about 15 days at a water temperature of 25° C. They still return to the female 's mouth for a few days at any sign of danger and overnight.

Pseudocrenilabrus philander philander (Weber, 1897)

Explanation of the scientific name: Refers to the association of the mates during the spawning

Pseudocrenilabrus multicolor during their spawning. The female has laid some eggs, the male has fertilized them and is now flashing his anal fin over the eggs so the female can pick them up and, perhaps, get an extra serving of sperm to fertilize those eggs which she picked up before fertilization.

The female has her throat patch quite distended with eggs she has gathered, but she keeps on spawning.

period.

philandros (Gk.) = loving one's spouse.

Other names in the literature:
Chromis philander Weber, 1897
Paratilapia moffati Boulenger, 1898 (non Castelnau, 1861)
Tilapia philander Boulenger,1899 (in part)
Tilapia ovalis Boulenger, 1901 (non Steindachner, 1866)
Haplochromis moffati Boulenger, 1907
Astatotilapia moffati Stoye, 1932
Pseudocrenilabrus natalensis Fowler, 1934
Hemihaplochromis philander Wickler, 1963.

Original description: Weber, M. (1897): Zur Kenntnis der Susswasserfauna von Sudafrika. *Zool. Jahrb. Abt. Syst. Geogr. Biol.,* vol. 10, p. 148 (*Chromis philander*).

Distribution: Southern Africa, north as far as the region of the source of the Congo (Zaire).

Habitat: Rivers and lakes, in places with dense plant growth.

First importation into Germany: This species has not yet been imported alive.

Characters:
fins: D XIII-XV/9–11, A III/8–10
total length: male up to 100 millimeters; female up to 110 millimeters.

Sex differences: Males are more colorful.

Pseudocrenilabrus philander dispersus (Trewavas, 1936)
Copper Mouthbrooder

Explanation of the scientific name: Refers to the large area of occurrence.
disperse (L.) = scattered, occurring here and there.

Other names in the literature:
Haplochromis philander dispersus Trewavas, 1936
Hemihaplochromis philander dispersus Wickler, 1963.

Original description:
Trewavas, E. (1936): *Novitates Zoologicae. Tring. Mus.* 40, p. 73 (*Haplochromis philander dispersus*).

Distribution: Southern Africa, north as far as Mozambique.

Habitat: Marshy regions and standing to slow-flowing bodies of water with heavy plant growth.

First importation into Germany: 1911 by the Vereinigten Zierfischzuchtereien Conradshohe (Berlin).

Characters:
fins: D XIV-XV/9–11, A III-IV/8–10

scales: mLR 26–28

total length: male up to 100 millimeters; female up to 110 millimeters.

Sex differences: Males are more colorful. The bright red spot at the end of the anal fin is conspicuous.

Aquarium keeping: As in *Pseudocrenilabrus multicolor.*

Spawning: As in *Pseudocrenilabrus multicolor.*

The female doesn't eat while she carries the developing eggs and subsequent fry in her mouth. Her body begins to shrink in size, especially the area around the caudal peduncle. (Below) A male *Pseudocrenilabrus philander dispersus* presenting his spawning dance on a special spawning site, hoping to attract a ripe female.

Pseudocrenilabrus multicolor.

Pseudocrenilabrus philander dispersus.

Pseudocrenilabrus philander dispersus, head study.

Pseudocrenilabrus philander dispersus, ♀ with brood in throat sac.

Tanganicodus irsacae.

This photo of _Haplochromis burtoni_ clearly shows the egg spots on the anal fin of the male mouthbrooder. The theory is that the female, thinking these are more eggs, snaps at them. At the same time the male is supposed to release sperm which the female gets into her mouth to fertilize those eggs which she picked up prior to their being fertilized by the male.

Genus *Pseudotropheus* Regan, 1921

Explanation of the scientific name: Refers to the separation from the genus *Tropheus*.

pseudo (L.) = false; *Tropheus* = cichlid genus from Lake Tanganyika, established by Boulenger in 1898.

Original description: Regan, C. T. (1921): The Cichlid fishes of Lake Nyasa. *Proc. Zool. Soc. London,* p. 681.

Distribution: Lake Malawi (Africa).

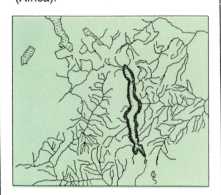

Number of species: So far about 25 species are known.

Generic characters: The dentition consists of an outer row of bicuspid teeth and of additional rows of tricuspid teeth located farther to the inside. This, however, is also true of the genus *Melanochromis*. The genus *Pseudotropheus* differs from the genus *Melanochromis* on the basis of the larger number of phyrangeal teeth. In addition, the pharyngeal teeth are smaller than in *Melanochromis*. The dorsal fin has 16 to 19 spines and 8 to 10 soft rays. The anal fin has three spines and 7 to 9 soft rays. There is clear sexual dimorphism.

Total length: 70 to 150 millimeters.

Type species: *Pseudotropheus williamsi* (*Chromis williamsi* Günther, 1893).

Pseudotropheus aurora Burgess, 1976

Explanation of the scientific name: Refers to the light golden-yellow color of the breast and lower head region, which is particularly conspicuous on account of the bluish ground color of the body.

aurora (L.) = dawn, sunrise.

Other name in the literature: *Pseudotropheus lucerna*?

Original description: Burgess, W. E. (1976): *Pseudotropheus aurora*, a new species of cichlid fish from Lake Malawi. *Tropical Fish Hobbyist,* vol. 24, nr. 9, p. 52–56.

Distribution: Close to shore at Likoma Island (Lake Malawi).

Habitat: Transition zone from rocky to sandy substrate.

First importation into Germany: 1977.

Characters:

fins: D XVII-XVIII/9–10, A III/7–8, P 13–14

scales: mLR 29–33

total length: male up to 110 millimeters; female up to 100 millimeters.

Sex differences: Males are blue and yellow colored; females are brown-gray.

Aquarium keeping: It has proved effective, if possible, to keep the fish in large tanks (at least 130 centimeters long) with several other Malawi cichlids. One should also build numerous hiding places for them in the form of fairly large rock structures. A planting with various tough-leaved aquatic plants is possible. As a substrate one uses gravel with a grain size of from 2 to 5 millimeters. The fish make no particular demands with respect to water hardness; however, the *p*H should not fall below the neutral value of 7. The water temperature can be between 24° and 28° C. Besides feeding them all available kinds of live food, one should also offer the fish dry food containing a large percentage of vegetable matter.

Spawning: The fish spawn in the community tank without difficulty. One can recognize the ready-to-spawn female by the fuller belly area and by the fact that the spawning tube protrudes by up to 1 millimeter. After a brief courtship, which, however, is somewhat more than guide swimming to the spawning site, the mates abruptly begin to circle each other, in the process of which they mutually nudge the anal region or the anal fin with the mouth. In this manner an egg is released each time and is immediately taken by the female into her mouth after each turn. After that the female takes the sperm into her mouth at the male's genital region. In so doing she touches the egg spot on the male's anal fin. Fertilization apparently takes place in the female's mouth. Up to 40 eggs are laid. After all of the eggs are laid, several more mock matings are carried out, in which the female apparently takes up additional sperm. After spawning, the female swims with full throat sac to a sheltered site. The eggs are turned over repeatedly through constant chewing movements. The 10- to 14-millimeter-long fry are released from the mouth after about 24 days. During the time the eggs are held in the mouth, the female also takes in some nourishment. After they are released from the mouth, the fry continue to be taken into the female's mouth for a few days at any sign of danger.

Pseudotropheus elongatus
Fryer, 1956
Elongate Cichlid

Explanation of the scientific name: Refers to the build of the fish.

elongatus (L.) = elongated.

Original description: Fryer, G. (1956): New species of Cichlid fishes from Lake Nyasa. *Rev. Zool. Bot. Afr.,* vol. 53 (1–2), p. 81–91 (*Pseudotropheus elongatus*).

Distribution: Lake Malawi (Nkhata Bay, Mbamba Bay).

Habitat: Boulder and rock zones with algal growth.

First importation into Germany: 1964.

Characters:

fins: D XVII-XXVIII/8–9, A III/7

total length: male up to 120 millimeters; female up to 110 millimeters.

Sex differences: Males are more intensely colored. The ground color is dark blue, almost black, with light blue to lilac vertical bars. The occiput is also light blue to lilac colored. Females are pigeon-blue to gray.

Aquarium keeping: As in *Pseudotropheus aurora.*

Spawning: As in *Pseudotropheus aurora.*

Genus *Spathodus* Boulenger, 1900

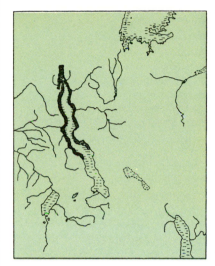

Explanation of the scientific name: Refers to the shape of the teeth.

spatha (L.) = broadsword, sword.

Original description: Boulenger, G. A. (1900): Matériaux pour la faune du Congo. Poissons nouveaux du Congo. *Ann. Mus. Congo, Zool.,* ser. 1, p. 152.

Distribution: Lake Tanganyika (Africa).

Number of species: So far two species are known.

Generic characters: The genus *Spathodus* is equipped with elongated gripping teeth that make it possible for the fish to extract and feed on microorganisms from patches of algae. Vertical bars are not present; they have two incomplete lateral lines.

Total length: 80 to 100 millimeters.

Type species: *Spathodus erythrodon* Boulenger, 1900.

Comments: The fishes of the genus *Spathodus,* along with those of the genera *Eretmodus* and *Tanganicodus,* belong to the so-called Goby Cichlids (tribe Eremodini).

Spathodus erythrodon
Boulenger, 1900
Blue-Spotted Gray Cichlid

Explanation of the scientific name: Refers to the brownish to reddish coloration of the fish.

erythros (Gk.) = red, reddish.

Original description: Boulenger, G. A. (1900): Matériaux pour la faune du Congo. Poissons nouveaux du Congo. *Ann. Mus. Congo, Zool.,* ser. 1, p. 152 (*Spathodus erythrodon*).

Distribution: Lake Tanganyika.

Habitat: Shallow water zones, usually in regions with sand or boulders, where they principally feed on algae and the microorganisms living in them.

First importation into Germany: 1958.

Characters:

fins: D XXIII/5, A III/6–7

total length: male up to 85 millimeters; female up to 80 millimeters.

Sex differences: No clearly visible external characters.

Aquarium keeping: One keeps the fish in pairs in large aquaria, if possible, but also in community tanks with other Tanganyika cichlids. In the furnishing, in addition to using fine-grained gravel (grain size of about 2 millimeters), one should make sure that numerous hiding places in the form of cave-like rock structures are present. The background can be planted with *Bolbitis heudelotti.* The fish are fed *Cyclops* and dry food with vegetable components. The water temperature should be between 23° and 28° C.

Spawning: Pair formation already begins several days before spawning. One then observes the fish together constantly. During this time the male and female court each other by turns again and again. For spawning, the pair searches for as smooth a rock as possible; somewhat diagonally sloping, almost level rocks are preferred. In the spawning process the fish circle each other, during which they mutually touch the anal region with the mouth. The eggs released by the female, which are discharged individually or in pairs, are immediately taken into the female's mouth. Fertilization occurs in the female's mouth. At the end of spawning, in which more than 25 eggs can be released, the female broods the eggs in her mouth and transfers the larvae to the male about 11 or 12 days after spawning. The male keeps them in its mouth up to the twenty-first day. The free-swimming, about 10-millimeter-long fry seek out sheltered places in the aquarium immediately after they are let out of the mouth, and then are no longer cared for by the parents. It is best to feed the fry with rotifers and *Artemia* nauplia.

Spathodus marlieri Poll 1950

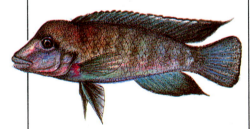

Explanation of the scientific name: Refers to the Belgian ichthyologist G. Marlier.

Original description: Poll, M. (1950): Histoire du peuplement et origine des espèces de la faune echthyologique du Lac Tanganyika. *Ann. Soc. Royal Zool. Belg.*, p. 111–140 (*Spathodus marlieri*).

Distribution: Northern part of Lake Tanganyika.

Habitat: Boulder zones near shore.

First importation into Germany: 1975.

Characters:
fins: D XXI/6–8, A III/7
scales: mLR 30–31
total length: male up to 100 millimeters; female up to 90 millimeters.

Sex differences: Older males exhibit a humped forehead; dorsal and anal fins are tapered to a point.

Aquarium keeping: As in *Spathodus erythrodon*.

Spawning: As in *Spathodus erythrodon*, except that the larvae are not transferred from the female to the male.

Genus *Tanganicodus* Poll, 1950

Explanation of the scientific name: Refers to the area of occurrence, Lake Tanganyika.

Original description: Poll, M. (1950): *Tanganicodus irsacae*, gen. n., sp. n. *Rev. Zool. Bot. Afr.*, vol. 43, 4, p. 297.

Distribution: Lake Tanganyika (Africa).

Number of species: So far only one species is known.

Generic characters: In comparison to the other Goby Cichlid genera (*Eretmodus* and *Spathodus*), the genus *Tanganicodus* possesses a smaller mouth with the upper mandible clearly projecting beyond the lower. The central teeth of the single, irregular row of teeth are strikingly long and pointed.

Total length: Up to 70 millimeters.

Type species: *Tanganicodus irsacae* Poll, 1950.

Tanganicodus irsacae Poll, 1950

Explanation of the scientific name: Refers to the proper name Irsac.

Original description: Poll, M. (1950): *Tanganicodus irsacae*, gen. n., sp. n. *Rev. Zool. Bot. Afr.*, vol. 43, 4, p. 297.

Distribution: Northern part of Lake Tanganyika.

Habitat: Boulder zones near shore, where the fish feed on algae growing on the rocks and the microorganisms living in it.

First importation into Germany: 1975.

Characters:
fins: D XXIII-XXIV/4–5, A III/7
scales: mLR 30–32
total length: both sexes up to 70 millimeters.

Sex differences: No visible external characters.

Aquarium keeping: As in *Spathodus erythrodon*.

Spawning: As in *Spathodus erythrodon*, except that the larvae are not transferred from the female to the male.

**Genus *Triglachromis*
Poll and Thys, 1974**

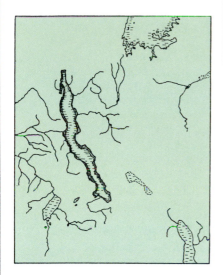

Spathodus erythrodon

Explanation of the scientific name: Refers to the marine fish family Triglidae, the members of which are also called Gurnards or sea robins, and to the former name for cichlids, "chromides."
chroma (Gk.) = color.
Original description: Poll, M., and Thys van den Audenaerde, D. F. E. (1974): Genre nouveau *Triglachromis* propose pour *Limnochromis otostigma* Regan, Cichlidae du Lac Tanganyika. *Rev. Zool. Bot. Afr.,* vol. 88, p. 127–130.
Distribution: Lake Tanganyika (Africa).
Number of species: So far one species is known.
Generic characters: The genus *Triglachromis* is closely related to the genus *Limnochromis*, from which it was removed by Poll and Thys van den Audenaerde (1974). Differences are evident in the form of the mouth and the pectoral fins. The base of the pectoral fin is much lower than in *Limnochromis* species. It begins almost at the side of the belly. The individual rays of the pectoral fin, particularly the middle ones, are elongated beyond the membranes. They are curved downward and look like claws. This is the reason for the comparison with the gurnards or sea robins.

Total length: 120 millimeters.
Type species: *Triglachromis otostigma* (*Limnochromis otostigma* Regan, 1920).

Triglachromis otostigma (Regan, 1920)
Tanganyika Gurnard
Explanation of the scientific name: Refers to the dark spot on the gill cover.
stigma (L.) = mark, brand; *oto-* (L.) = pertaining to the ear.
Other name in the literature:
Limnochromis otostigma Regan, 1920.
Original description: Regan, C. T. (1920): The classification of the fishes of the family Cichlidae. I. The Tanganyika genera. *Ann. Mag. Nat. Hist.,* vol. 5, p. 43 (*Limnochromis otostigma*).
Distribution: Lake Tanganyika.
Habitat: At depths of 10 to 50 meters on muddy bottoms.
First importation into Germany: 1973.
Characters:
fins: D XV-XVI/8, A III/8

total length: male up to 120 millimeters; female up to 100 millimeters.

Sex differences: Males have greatly elongated ventral fins and a rich blue coloration.

Aquarium keeping: The tank size should not be less than 80 centimeters. Fine-grained gravel (grain size about 2 millimeters) serves as the substrate. One provides the fish with good hiding places by furnishing rock structures with numerous caves. The aquarium can be planted with *Bolbitis heudelotti* or Java Fern. The water temperature should be between 23 and 28° C.

EXPLANATION OF TECHNICAL TERMS

Adhesive gland: Gland on the head from which a sticky secretion is released, with which the larvae can attach themselves to substrates.

Affinis: (Lat.) (abbr.: aff.): related, similar.

Albino: individual lacking pigmentation; whitish with red eyes; hereditary metabolic disorder.

Artemia salina: Brine shrimp. Species of marine shrimp that principally comes from the Great Salt Lake in Utah, the eggs of which are offered in the aquarium trade. These eggs can be stored in sealed containers for a long time. The larvae hatch in warm salt water and are a good first food for fry.

Biotope: Living space with specific environmental conditions, in which characteristic animal and plant species find the conditions necessary for their existence.

Black water: Soft, acidic water enriched with humic acids, with a brownish color; pH of 3.8 to 4.5.

Brood care: The totality of mainly innate parental behavior patterns that serve the protection and rearing of the offspring.

Buffering: The effect of specific substances on the stabilization of the pH.

Carbonate hardness: The amount of magnesium and calcium hydrogen carbonates dissolved in the water; also see total hardness.

Clear water: Very clear fresh water with great range of sight, often with a slight greenish tint; as a rule, the pH lies between 4.6 and 6.6.

Communication: Mutual understanding, exchange of information.

Courtship: Specific innate behavior patterns, which indicate the readiness to reproduce.

Cyclops: Crustaceans of the genus Cyclops; good rearing food.

Daphnia: Water fleas; crustaceans of the genus Daphnia.

Dry food: Collective name for industrially produced food for exotic fishes.

Ecological niche: Position of a species in the ecosystem; for example, as a link in a food chain. Where the species finds the conditions required for its existence.

Ecosystem: Totality of the animate and inanimate elements of an area of the biosphere in their interrelationships and dependencies.

Ecotope: The smallest ecological unit of volume of a landscape with uniform plant and animal resources.

Fertilization: Union of maternal and paternal genes through the fusion of the nuclei of the egg and sperm cells.

Family: Systematic main category that is placed above the subfamily or genus and below the superfamily or order.

Fin base: Line of attachment of the fin on the body (external).

Form (L. forma): Shape, appearance; any tangible taxonomic deviation from the type of the species, although not classifiable as a species or subspecies.

Free-swimming: The ability to swim achieved by the fry at the end of the larval stage.

Genus: Systematic category, placed under a family or subfamily, usually consisting of several closely related species.

Genital: Referring to the sex organs (genitalia).

Genital papilla: Protrusion at the male's genital opening.

Habitus: External appearance, phenotype.

Holotype: Specimen used for the determination. The single specimen selected by the author of the original description, upon which the description and naming of the new species or subspecies are based. Additional specimens available for the original description are called paratypes.

Ichthyology: The study of fishes.

Igarapé: Brazilian name for natural canal-like connections between two bodies of water.

Individual: A single specimen of a species.

Intimidation: Display of physical dominance in courtship and aggressive behavior; for example, through displaying a conspicuous coloration and through particular movements and postures, such as fin spreading.

Larva: Developmental stage after the hatching of the young fish, in which the swim bladder is not yet functional and nourishment is still taken from the yolk sac.

Lateral: On the side.

Lateral band: Side band, longitudinal band; as a rule, a dark stripe along the sides of the body.

Lateral spot: A spot located on each side of the body.

Lateral line: Sense organ of fishes for the detection of the slightest water movements, developed on the sides of the body as a continuous or divided line.

Longitudinal stripe: See lateral band.

Mutant: Individual with new inheritable traits altered by mutation.

Nomenclature: The system of scientific names.

Pharyngeal teeth: Teeth standing in one to three rows on the lower phyrangeal bone; they may be long and pointed, but also cobble-stone-like.

Popular name: Common or vernacular name.

Population: Totality of individuals with the same genetic traits present in a continuous area.

Releaser: Stimulus or combination of stimuli; triggers a specific behavior pattern.

River basin: Region of the immediate course of a river, including the mouths of the tributaries.

River system: Total drainage area of a river; includes all tributaries and bodies of water in the region.

Sexual dichromatism: Difference in coloration between the sexes.

Sexual dimorphism: Difference in shape or build between the sexes.

Spawning papilla: Protrusion at the male's or female's genital opening.

Spawning substrate: Base on which the eggs are laid, such as plants, rocks, or artificial materials.

Sperm: Totality of sperm and seminal fluid.

Synonym: (Gk. *synonyma*) (abbr.: syn.): A former scientific name that is no longer valid in the nomenclature.

Systematics: (Syn.: taxonomy): Classification of organisms through description, naming, and arranging in a system according to their phylogenetic relationships.

Territory: The space required by an individual or a pair.

Total hardness: Totality of all of the calcium and magnesium salts dissolved in water; as a rule, the sum of the carbonate and noncarbonate hardness.

Type species: The species that is considered typical of a genus.

Type specimen: The single specimen selected as the type for the description of a new species (see holotype).

Variety (abbr.: var.): Subspecies, race; group of individuals with the same inheritable characters that differ from the type species. Varieties of a species can interbreed with one another without restriction.

White water: Water with a high content of suspended mineral matter that is carried along from the mountainous drainage area, in South America, for example, from the Andes; pH between 6.5 and 7.

Whiteworms: Whitish, about 30-millimeter-long segmented worms that live, among other places, in compost heaps and are raised as food for the aquarium hobby.

Xanthorism: Gold coloration; inheritable deviant coloration.

Xanthistic forms: Individuals with a yellowish-gold coloration; in contrast to albinos, they do not have red eyes.

AFRICAN CHICLIDS OF LAKES MALAWI AND TANGANYIKA
Twelfth Edition
By Dr. Herbert R. Axelrod and Dr. Warren E. Burgess
ISBN 0-87666-792-2
TFH PS-703

The newest edition of this best-seller is completely redesigned so that the reader will be able to more easily find and identify his fishes.

In addition, the text has been completely rewritten to bring the book more up-to-date with the hobby as it has progressed since the first edition. More information has been added in our usual easy-to-read style.

There are additional photos of species not seen before, the names have been brought up to date if changes have occurred since the last edition, and the master checklists of cichlids of Lakes Malawi and Tanganyika have also been added to or modified in accordance with new species described or changes in the old species.

Hard cover, 5½ x 8', 448 pages. Over 550 full-color photos.

CICHLIDS OF THE WORLD
By Dr. Robert J. Goldstein
ISBN 0-87666-032-4
TFH H-945

A comprehensive book on the most popular group of fishes. For the avid aquarist who specializes or wants to specialize in cichlids. A must for the reference library of advanced aquarists, dealers, fish importers.

Hard cover, 5½ x 8', 382 pages. 104 black and white photos, 270 color photos

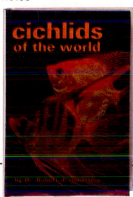

FISHES OF LAKE TANGANYIKA
By Pierre Brichard
ISBN 0-87666-464-8
TFH H-972

This handsome big book, highly il-lustrated with excellent color pho-tos of the fishes of Lake Tangan-yika and of the lake and its environs, is designed to be an im-portant reference work for aquar-ium hobbyists specializing in Afri-can cichlids, for commercial suppliers to the aquarium trade, for ichthyologists and students of ichthyology. Covers all aspects of aquarium management and breeding as well as taxonomy and ecology.

Hard cover, 5½ x 8½", 448 pages. 84 black and white photographs; 285 color photos; 58 line illus-trations.

EXOTIC TROPICAL FISHES EXPANDED EDITION
By Dr. Herbert R. Axelrod, Dr. Cliff W. Emmens, Dr. Warren E. Burgess, Neal Pronek and Glen Axelrod
Permanent hardcover binding
ISBN 0-8766-543-1
TFH H-1028
Looseleaf hardcover binding; 2 vols.
ISBN 0-8766-537-7
TFH H-1028L

Aquarium Management—Mainte-nance and management of aquaria and aquarium systems (128 pages). Includes information on foods and feeding, breeding, preventing, recognizing and curing of diseases.

Exotic Aquarium Plants—Covers all aspects of aquarium plant management from planting to plant cultivation (128 pages).

Raising Tropical Fishes Commercially—Setting up and management of indoor and outdoor hatcheries (32 pages).

Descriptive Catalog of Exotic Tropical Fishes—Provides 960 pages of the most up-to-date information on the water, food, and environmental requirements of nearly every fish that's ever been kept in hobbyists' tanks. Over 1,000 full-color photos. Expanded edition includes all the information that was provided in the first 25 supplement books to EXOTIC TROPICAL FISHES.

Hardbound and looseleaf, 5½ x 8½", 1,312 pages.
(hardbound, non-looseleaf) 1 vol.
(looseleaf) 2 vol. set.

DR. AXELROD'S ATLAS OF FRESHWATER AQUARIUM FISHES
Second Edition
By Dr. Herbert R. Axelrod and others
ISBN 0-86622-139-5
TFH H-1077

Here is a new book—a truly beautiful and immensely colorful new book—that satisfies the long existing need for a *comprehensive* identification guide to aquarium fishes that find their way onto world markets. This book describes AND SHOWS IN FULL COLOR not only the popular aquarium fishes but also the oddballs and weirdos, not just the warmwater species but the coldwater species as well, not just the foreign fishes but the domestic species too.

But not all of this beautiful text/photo package is concerned with identification and maintenance alone. In addition to showing the fishes and telling exactly what they are, the book also provides information—plus step-by-step full-color photographic sequences—about the spawning of a number of species from different families.

Illustrated with more than 4000 full-color photos.
Hard cover, 8½ x 11", 784 pages.

DR. AXELROD'S MINI-ATLAS OF FRESHWATER AQUARIUMS
By Dr. Herbert R. Axelrod, Dr. Warren E. Burgess, Dr. Cliff W. Emmems, et. al
ISBN 0-86622-385-1
TFH H-1090

The entire staff of TROPICAL FISH HOBBYIST magazine, aided by Prof. C. W. Emmens, pooled their talents to make this the most complete book on aquarium fishes ever published.

This is called the "mini" ATLAS in reference to the much larger DR. AXELROD'S ATLAS, which, in its last edition, had more than 4,000 color photographs of fishes, but which lacked the selection, care, breeding, and aquarium plants material. The new ATLAS includes a full-length section treating the most up-to-date methods of aquarium fishes and plant care by one of the world's leading authorities, Dr. Cliff Emmens. The scientific names of the fishes are up-to-date with the most recent, scientifically accepted designations. The text and photos in the new MINI-ATLAS has a separate value to hobbyists: all of the fishes in the book are cross-referenced by both scientific and common names, making it possible to obtain information about a species even if you don't know its scientific name.

A big FIRST for this book is the pictorial captions, which tell at a glance all the pertinent information the average person needs to know about the fish . . . how it breeds, its temperature, feeding habits, etc. In that way the authors have been able to present a huge amount of information in the smallest possible space.

Hard cover, 5½ x 8½", 992 pages. Over 2200 full-color photos.

DWARF CICHLIDS
By Dr. Jörg Vierke
ISBN 0-86622-982-5
TFH TS-118

Sharing in all of the good points—colorfulness, ease of keeping and breeding, fascinating behavior—that have endeared their larger cichlid cousins to aquarists all over the world, dwarf cichlids have some advantages over the big boys. Because they're smaller they are more comfortable in smaller tanks, making them more suitable for keeping by beginners, and in general they're also less aggressive than the larger cichlids. This excellent book provides vital information about all of the dwarf cichlids—how to set up their tanks, how to breed them, how to keep them healthy—all accompanied by full-color photos and drawings that serve as a perfect identification guide. This book is revealing and informative even for advanced aquarists, and it's a treasure-trove of valuable tips and insights for the less experienced.

Hard cover, 5½ x 8', 160 pages. 191 full-color photos and drawings

Page numbers printed in **bold** refer to photgraphs or illustrations